The Power of Deduction

The Power of Deduction

Failure Modes and Effects Analysis for Design

Michael A. Anleitner

ASQ Quality Press
Milwaukee, Wisconsin

American Society for Quality, Quality Press, Milwaukee 53203
© 2011 by ASQ
All rights reserved. Published 2010
Printed in the United States of America
16 15 14 13 12 11 10   5 4 3 2 1

Library of Congress Cataloging-in-Publication Data

Anleitner, Michael A.
    The power of deduction : failure modes and effects analysis for design /
Michael A. Anleitner.
        p. cm.
    Includes bibliographical references and index.
    ISBN 978-0-87389-796-9 (hardcover : alk. paper)
    1. System failures (Engineering)   I. Title.

    TA169.5.A58 2010
    620'.00452—dc22   2010027597

ISBN: 978-0-87389-796-9

Publisher: William A. Tony
Acquisitions Editor: Matt T. Meinholz
Project Editor: Paul O'Mara
Production Administrator: Randall Benson

ASQ Mission: The American Society for Quality advances individual, organizational,
and community excellence worldwide through learning, quality improvement, and
knowledge exchange.

Attention Bookstores, Wholesalers, Schools, and Corporations: ASQ Quality Press
books, video, audio, and software are available at quantity discounts with bulk
purchases for business, educational, or instructional use. For information, please
contact ASQ Quality Press at 800-248-1946, or write to ASQ Quality Press,
P.O. Box 3005, Milwaukee, WI 53201-3005.

To place orders or to request ASQ membership information, call 800-248-1946. Visit our
Web site at http://www.asq.org/quality-press.

 Printed on acid-free paper

Quality Press
600 N. Plankinton Avenue
Milwaukee, Wisconsin 53203
Call toll free 800-248-1946
Fax 414-272-1734
www.asq.org
http://www.asq.org/quality-press
http://standardsgroup.asq.org
E-mail: authors@asq.org

Table of Contents

List of Figures and Tables

Preface

I've been trying to write this book for almost 10 years. I've started and quit at least four times before (I've lost track, actually) and created more than a dozen outlines that have never worked out for one reason or another.

I've tried to write about *failure modes and effects analysis* (FMEA) using different formats, too. I've tried using traditional prose and I've also tried to write this book using graphics as a primary feature. I even gave serious consideration to creating a "pop-up" book of the type used to educate toddlers.

No matter the approach, my earlier efforts consistently collapsed.

There are a lot of reasons for this. Twice, my efforts were derailed by professional and personal challenges in my life. There's only so much time in every day, and writing this book never took priority over my obligations to my family or my work with clients.

Most of the time, though, I became discouraged about what I was trying to write. The subject can be so complicated and intricate that creating clear and unambiguous explanations for some of the most important concepts in FMEA can be torturous.

On more than one occasion I've spent an entire day writing only two or three pages. After a while, that kind of exertion can drain your creative spirit for several days or even several weeks.

The final reason I've had trouble sticking with this project, though, has been more difficult to overcome. Virtually each time I facilitated or directly worked on a *design FMEA* (DFMEA) study, I had one or more significant "Aha!" moments. Every time I learned something new, I revised the teaching materials I was using as a facilitator, and I've kept an entire archive of presentation files going back to the mid 1990s. The progression of understanding visible in this archive is not trivial.

I started to think I would never be able to write this book because, in all honesty, I wasn't sure that I understood FMEA well enough to explain everything about it in a clear and coherent way. In the past two years, however, I found that the "pop–pop–pop" of new ideas I had seen in every project tapered off, and I had fewer and fewer major insights.

And, even when I seemed to get the technical part of it correct, I found that it was a challenge to write something that would be entertaining, too.

That's one of my goals here as well. I hope that you will enjoy this book because I've discovered that adults learn much more when they are entertained and delighted while exploring new ideas.

To get all of this down on paper, I tried to be as general as possible. However, FMEA, particularly DFMEA, has been used more in the automotive industry than anywhere else. Plus, more than half of my FMEA experience has been in automotive project work. As a result, there are many, many automotive examples in the book.

The automotive industry is possibly the most studied industry in the world, with the most public information, and it's probably the most competitive legal business in the world, too. (Many illegal businesses are a bit more competitive.) That means there are plenty of well-understood examples that most people reading this book might know about. And most engineers either love cars or at least have strong opinions about cars. So, automotive examples form the backbone of this book.

In this same vein, mechanical systems have been subjected to more DFMEA studies than electronic systems, software, or services. Many people, even those who don't have strong backgrounds in the underlying physics, can understand mechanical systems. In addition, a good deal of automotive technology is based on mechanical engineering. For these reasons I've used mechanical systems to illustrate most of the major points in the book.

While doing this, I've tried to show that these same principles can be applied to nonmechanical projects. I will fully admit, though, that there are twists and turns that will arise in nonmechanical studies that I haven't addressed.

I must also acknowledge that my role as an engineer is one of practitioner, not theoretician. Although underlying theory plays a major part in this book, and I've tried to be reasonably thorough in an academic sense, I haven't tried to write something that could be accepted as a peer-reviewed paper. The topic is too long and involved for any journal, and ultimately I wasn't really interested in that level of formalism. Had I tried to do that, I would never have completed this endeavor.

When you tabulate the skull sweat and occasional tears that went into this book—and the kind of return that an author can expect for writing about a specialized topic like FMEA—you'll realize that no one in their right mind does this sort of work for the money.

I did this to help people be more productive in creating new products. By improving products, I firmly believe that we can work together, across national boundaries, to build greater wealth for people throughout the world. By building wealth, we aid the quest for world peace by bringing greater prosperity to more and more of the world. When prosperity is widespread, humanity is less likely to engage in threats, violence, and warfare.

I hope that this book plays a small part in bringing these benefits to everyone on the planet.

Michael A. Anleitner
May, 2010

1

Important Ideas About DFMEA

SOME BASIC VOCABULARY

It's often true—quality geeks explore the smallest and seemingly most trivial details.

Probing questions often turn into debates—and these debates sometimes morph into arguments. Arguments about fundamental issues are often passionate and occasionally significant. One of the most important areas that quality geeks debate is what might be called *core terminology*. Even a very simple discussion requires explanations for fundamental terms. We'll consider formal definitions for important expressions throughout this book, but for now we just need to describe a bit of the general vocabulary and acronyms that affect our subject.

At this point, there's no need to get worked up about jargon, but in this book I'll be talking about *failure modes and effects analysis* or FMEA. Engineers often pronounce this as "*fee*-ma," but that does run the risk of confusing our subject with the U.S. Federal Emergency Management Administration, or FEMA. Of course, it's likely that FEMA could use a lot of FMEA, but that's the kind of thing that can make your head spin.

In reality, FMEA is a formal process or study in which a subject is examined in detail and risk is assessed. A *design FMEA* (DFMEA) looks at risk from a product design perspective. A DFMEA could be about a component with a few parts, an assembly that includes several components, or even an indivisible piece of hardware, like a one-piece brake rotor, a capacitor, or even a simple bolt.

A *system FMEA* (SFMEA) is usually a high-level DFMEA encompassing several subassemblies and possibly dozens of components as well. As an example, an SFMEA could be conducted to understand how the major elements of an air conditioner—the motor, compressor, evaporator, condenser, refrigerant, and enclosure subsystem—would work together. However, an SFMEA usually considers the interactions of the major elements, not the details of how each subassembly or subsystem performs.

A *process FMEA,* or PFMEA, is a formal investigation of both fabrication and assembly processes. PFMEA can also be performed on any service process—even things as complicated as life-or-death surgical procedures—although there must be sufficient structure in a process for a PFMEA study to be worthwhile.

DFMEA techniques can also be used to explore the reliability of the design of a service process, such as a medication dispensing process in a hospital or the provision of table service in a restaurant. However, the *execution* of the process would be addressed with PFMEA methods.

So, let's start with a basic operational definition for FMEA—a definition that applies to any type of FMEA study.

FMEA is a structured, methodical technique. When done well, FMEA methodology is based on a repeatable set of steps or activities—in other words, a process.

The goal of any FMEA study is to identify what might go wrong before an error is actually made, whether that is an error in design or in realization of the design. Thus, a systematic approach is used to assess these potential errors in order to quantitatively prioritize risk.

In very simple terms, FMEA looks at what might go wrong, how bad this might be, how likely this undesirable event is, and how it might be either prevented or, alternatively, detected at the earliest possible moment—presumably before a customer might experience negative consequences.

At the most basic level, FMEA is a powerful technique that can increase customer satisfaction by preventing failures. For DFMEA, that goal can be restated as a systematic effort to eliminate design-based defects before the release of drawings, specifications, and plans for manufacturing, assembly, or construction of a product.

Similarly, PFMEA addresses manufacturing, construction, or service process issues with a goal of preventing manufacturing- or construction-driven failures, or execution errors for service processes.

Further, risk management includes finding, quarantining, and correcting errors once they are made. However, risk is lowered whenever prevention activities—as opposed to detection activities—are given more weight. Prevention controls can be far more powerful than detection controls and comprise an essential element of all world-class FMEA studies.

One of the most important things that you need to keep in mind, though, is that FMEA *can not eliminate risk.* It can tell you a great deal about risk and create semi-quantitative information or data about risk. But risk can not be eliminated, and decisions about risk must still be made. This can lead to serious ethical questions that FMEA can not—and should not—be asked to resolve directly.

FMEA, though, does make these discussions more fact-driven and less irrational. Nevertheless, some products are inherently risky. An automotive airbag system is a life-or-death product, and it is impossible to make an airbag system that has no flaws or weaknesses. We'll return to this issue of the ethical implications of FMEA in Chapter 10.

A BIT OF HISTORY

The first systematic attempt to institute a process for analyzing potential problems in product design appeared in late 1949, when the U.S. Department of Defense issued a military standard: MIL-P-1629. With the bulky title of "Procedures for Performing a Failure Mode, Effects and Criticality Analysis," or FMECA, this standard codified ideas and methods that had been applied less formally—and usually incompletely—for more than a century.

This standard was revised several times, with the final release in 1980. The procedure outlined in this standard was fully focused on design—as opposed to manufacturing process issues—and the scope was broad indeed, with the goal of using FMECA to not only guide design issues, but to affect maintainability, safety, survivability, logistics, and field problem diagnosis. Moreover, it was clearly targeted as system-level analysis, although subsystem and component-level design analyses were envisioned as possible. Software was specifically excluded from consideration.

While reading this document can be tedious (the 1980 edition is 54 pages long and filled with the technical jargon of military reliability), the basic ideas are remarkably similar to the best practices that will be explored in this book. The fact that these ideas have, with meaningful but less-than-fundamental changes, survived to this day is strong evidence that this methodology is powerful, reasonable, and, most importantly, *useful.*

The FMEA methodology has continued to evolve in many ways. It played a critical role in the Apollo moon shot program and was used with some intensity after the Apollo I launchpad fire that claimed the lives of three astronauts. We'll return to the Apollo I story later when we discuss the use of FMEA in development programs.

Soon, another fire problem caused FMEA to attract attention. Ford's Pinto was developed in the late 1960s, while the Apollo program was in full bloom. The Pinto was a low-cost subcompact car that was designed with a specific goal, namely to stem the tide of small foreign cars that were beginning to make serious inroads into the Detroit "Big Three" market share.

The Pinto development program was born amid internal controversy at Ford—Ford's president, Semon "Bunkie" Knudsen, opposed the idea while rising star Lee Iacocca, fresh off the triumph of the Mustang, pushed hard for a subcompact product. Henry Ford II sided with Iacocca, Knudsen soon resigned, and Iacocca turned his energy toward an aggressive agenda for weight and cost—the goals evolved into a "$2000 price and 2000 pound weight" mantra, although exactly where this arose is lost in the murky archives of automotive history.[1]

Further, Ford made a decision to develop the Pinto using a compressed time frame. Again, the details of this compression are far from clear, but many of those involved remember being asked to complete the design in just over two years at a time when four years was normally required to develop and launch a new vehicle.

This scenario makes the Pinto program particularly relevant in the twenty-first century—an engineering program with very aggressive functional and cost goals, combined with a shortened or accelerated schedule and an acute level of management scrutiny.

One consequence of these system-level objectives and insistent project goals was the placement of the fuel tank combined with the design of the rear axle and rear bumper system. Without entering the controversial world of exactly what might occur in test crashes and real-world impacts, there's little doubt that too often, when the Pinto was struck in the rear, the filler neck would break loose from the tank or the tank would crumple against the axle housing, sometimes splitting open the tank. In either case, fuel would leak out. Any ignition source—a spark from the crash or a hot exhaust system—could ignite the fuel.

Dozens—or perhaps hundreds—of people were seriously injured or killed.[2] Moreover, Ford was repeatedly sued, suffered terrible publicity, and, in a volatile article in *Mother Jones,* was excoriated from coast to coast through the author's depiction of Ford as a company that had a callous disregard for human life.

While many have debated (and probably will continue to dispute) the technical, ethical, and business details of this story, one thing is clear: a

---

1. It's far from clear that this was Iacocca's personal target, but it's very likely that he pushed the vehicle in that direction. And, as anyone who worked in Ford's engineering group in those days knew, if Iacocca even suggested that something was important, ignoring his will could be a terminal career choice.

2. There is a great deal of controversy about the number of people who were injured or killed.

better and more systematic evaluation of the design—an evaluation that was based on more than just testing a few cars—would have, in all likelihood, brought the overall issue to the fore.

As a result, Ford began using FMEA techniques on a broad scale in the mid-1970s. This was the intensive launch point for FMEA, and DFMEA in particular. Following Ford's example, the rest of the auto industry has slowly but surely tagged along, and FMEA is now a fairly common requirement for most (but not all) automakers and major parts suppliers.

Today, DFMEA is applied in many industries—but in none as intensively or in as much detail as in the automotive industry. PFMEA is applied in even wider circles; there are even specialized PFMEA techniques for healthcare and hospital operations.

TYPICAL APPROACHES

For better or for worse, DFMEA has been in relatively wide use for more than thirty years. Unfortunately, a good deal—perhaps most—of this usage has been "for worse" as the technique has been either misapplied or carried out in a way that either omits major issues or distorts the assessment of risk.

The First Common Problem: Forcing DFMEA without Supporting the Method

The first and most serious problem is that DFMEA is carried out because some higher authority requires it—and never really uses the resulting information to good advantage. Ford has been a major culprit in this area, even though Ford's engineering and quality community was (and continues to be) a pioneer in applying and driving FMEA methods and usage.

After using FMEA at the vehicle level, Ford started to require DFMEA and PFMEA studies from all suppliers. Eventually, the use of DFMEA largely disappeared within Ford itself, as it became something that Ford suppliers would use—but not Ford's engineering community. As suppliers were forced to adopt DFMEA, they were almost always resentful because Ford was engaged in a "do as I say, not as I do" game, a situation that nearly always infuriates everyone on the receiving end.

Worse, Ford used DFMEA (and, even more so, PFMEA) as a blunt weapon to beat concessions from suppliers and, when things went wrong, to make sure that most of the blame for vehicle problems was squarely and almost completely placed on the supplier community.

I was personally involved in one of these situations in the late 1970s. The firm I worked for (a company long since acquired and forgotten through several mergers and acquisitions) was producing parking pawls, the component that locks an automatic transmission in "park" and prevents unintended vehicle motion. The heat treatment of these metallic components was quite challenging, and failure of the pawl could cause a major safety problem.

In the aftermath of Ford's problems with the Pinto, FMEA had become very important, particularly for any items relating to vehicle safety. As an added bit of anxiety, Ford was under severe pressure from the U.S. National Highway Traffic Safety Administration (NHTSA) due to a tendency of some Ford transmissions to jump out of "park" with the engine running. There was some speculation that Ford might be forced to recall more than ten million vehicles and repair the defect—a defect that wasn't even fully understood.[3]

All of this commotion made the parking pawl a point of serious concern. In the DFMEA study, we found that specifying the correct heat treatment of the pawl was critical—and, of course, this also made the execution of the heat treatment equally critical. Ford even had (and still has today) a special symbol for these issues, called the "inverted Delta" or ∇.

There were several things that could go wrong in the heat treatment process. One or more pawls could have no heat treatment at all (something that can easily happen without sophisticated material handling controls), the part could have a hardness value lower than specified, or the depth of hardening could be less than needed for durability.

To make sure that nothing would go wrong, Ford insisted that our company purchase a very expensive eddy current test machine. By applying a magnetic field to each and every parking pawl and sensing the swirling fields (or eddy currents) that result at material discontinuities—like hardness zones—it's possible to infer a great deal about the heat treatment of any steel part.

This sounds easy in concept, but doing this for the entire surface of a complex part like a pawl is not simple at all. In particular, with the technology available in 1979—computers were still in the mainframe stage—it really required a full-time physicist to keep the test machine properly calibrated. Even today, accomplishing the level of control that our 1979-vintage FMEA required using eddy current testing is not a simple matter, particularly when 100% inspection of a high-volume part is required.

---

3. There was never any indication that parking pawls were at the root of this problem, though.

Of course, sooner or later things went wrong. Several pawls that had not been heat treated at all were shipped to the Ford transmission plant, and when these gearboxes were installed in cars, the "park" function went horribly wrong. Either the car would go into "park" and the pawl would deform so that the transmission could not be shifted out of "park," or, less often, the pawl would yield and slip, causing the transmission to fall out of "park."

Worst of all, this defect was discovered just a few days after NHTSA had given Ford approval to simply provide a warning sticker to owners of the 10 million vehicles that might have "park" problems. And here we were, about to reopen this can of worms and do it in a way that might make NHTSA rethink their relatively modest "fix" order.

Of course, we *had* goofed—and it had happened at the worst possible time. Ten thousand new Crown Victoria, Mercury Marquis, and Lincoln Town Car vehicles were piled up in a quarantine lot, and their transmissions would have to be torn down and every parking pawl replaced. Repairing these vehicles would cost almost one million dollars, and it could have been even worse had NHTSA changed their mind about the quick fix they had just approved.

But what did Ford do? They used our DFMEA and PFMEA studies as a device to establish that the entire responsibility for this fiasco was *our* liability, not theirs. Even though we had not included the eddy current test in our original PFMEA, Ford had insisted on this. We had proposed using a conventional hardness tester as a control, which would find low-hardness parts or "soft" parts that hadn't been heat treated at all.

There were three drawbacks to our proposal. It was very time-consuming and difficult to do this for every part. Secondly, this technique wouldn't find parts where the hardness depth was insufficient, nor would it find "spotty" hardness that wasn't uniform. Finally, putting a small indentation in the parts, as hardness testers do, wasn't appealing to Ford—even though creating the indentation away from the critical pawl surface would have no impact on the part's durability and would never be seen unless the transmission was torn down to replace the pawl.

Once we wrote the eddy current test into the PFMEA, Ford took the position that it was completely and undeniably our problem. This meant that the cost to repair the transmissions in already-assembled vehicles was totally our company's responsibility.

Without question, we bore significant responsibility, and eventually we compensated Ford for the majority of the costs incurred. On the other hand, we felt this was, to some extent, grossly unfair as well because Ford had mandated a technique that was really too complex for practical use.

Using the FMEA system as a "wedge" to shunt all of the responsibility onto our shoulders caused us to be very wary about what we entered in all future FMEA documents and to vigorously resist every effort (even good faith, technically sound efforts) by Ford to address risk issues in our FMEA studies.

In different ways, with different concerns, Ford repeated this scenario hundreds of times with suppliers over the years. In fact, it's difficult to find engineers who work for suppliers that do business with Ford who can't tell you a horror story of this ilk.

In fairness to Ford, however, I have noticed that in recent years, it really *has* been trying to correct this failing. The problem is that memories are long and, even today, some of Ford's quality and engineering people literally bully their supply chain whenever the results of an FMEA study aren't to their liking. And they still don't conduct system-level FMEA studies internally—or, if they do, they won't share them with key suppliers.

That's not to say that Ford doesn't have an important, even overriding interest in the design and manufacture of parts that are used in their products. However, to use FMEA as a wedge to force changes that aren't economic, practical, or even useful defeats the entire purpose of the FMEA methodology.

Whenever usage is forced, or the particular answers that are derived in FMEA are coerced, the results can be anywhere from useless to counterproductive. For example, if the results are manipulated to be noncommittal, the result has little value. And if management doesn't look carefully, they can easily conclude that there's little or no risk when a significant consequence is real.

The same thing can occur when requirements are dictated by international standards (such as the ISO 9001 standard, which evolved into ISO/TS 16949 in the auto industry) or simply by management insisting on the use of FMEA in product development systems without understanding what FMEA can and should be used to accomplish.

Whenever this happens, FMEA will become a bit of a sham or even a time-consuming, energy-draining exercise in futility. Eventually, in the same way that bad currency drives good currency out of circulation,[4] bad DFMEA practices drive good DFMEA practices out of the engineering world.

This chain of events can even arise when there is no outside force for FMEA, such as a large customer who demands it or an industry or international standard that requires FMEA. It can even occur when executives think FMEA would be helpful but don't really understand the underlying methodology.

---

4. Economists call this principle *Gresham's law.*

The real key to avoiding this problem is for management to be fully supportive—and even a bit obsessive—about DFMEA. DFMEA is probably the most important tool or methodology that can be used to drive excellence in design. It's not the sole intellectual tool that can (and should) be used to achieve outstanding design results, but it may well be the most important.

Managers need to understand the DFMEA method and know what a DFMEA can—and can't—accomplish. Then, they need to review every DFMEA that they commission. After all, if it's important enough to ask working-level engineers to conduct the study, it's important enough for management to examine the results.

Finally, managers need to use DFMEA as a powerful discipline to prevent design errors. This is related to the idea that DFMEA results need to be reviewed by managers, but there's more to it than just reading, understanding, and asking constructive questions about the results of any DFMEA study.

Whenever managers adopt a continuing interest, an interest that is both sincere and has depth, FMEA studies improve significantly. More importantly, this interest reinforces and encourages continuing use of FMEA techniques throughout product development activities. Under these circumstances, the idea that DFMEA studies are forced disappears, and the utility and value of DFMEA studies increases dramatically.

The Second Common Problem: Adopting a "Form" Mentality

All FMEA studies are summarized on some kind of form. The Automotive Industry Action Group (AIAG) has several different forms that they recommend for the auto industry, and there are hundreds of formats in use today.

All of these forms are similar, and we'll look at the basic content of a typical DFMEA form repeatedly in the following chapters.

However, the use of a form to record the results may lead to another major difficulty. The completion of the worksheet itself, rather than active learning from the study, can easily become the most important goal in a product development process. This can often be a result of forcing DFMEA studies without support, but this can occur even when management backing appears strong and positive.

Because most DFMEA forms seem logical and are relatively easy to understand, they cause most people who work on DFMEA studies to assume that there's no real underlying technique needed to do a study. Just fill in the lines on the form and you'll be done.

I suppose it might be possible to develop a sound result just by following the apparent logic of whatever form you might be using. I've never seen
that happen, though, and I've been doing FMEA studies for more than thirty
years. In all that time, I have never—not once—seen an FMEA worth the
paper it was written on that was done by "just using the form."

There are a number of common symptoms that almost anyone can
recognize when a "form" mentality has been followed. Here are the items
I typically look for—you may see one, two, or even all of these in a poorly
executed study:

- There doesn't seem to be a comprehensive or clear delineation
  of the various levels of design in the DFMEA. As a result, the
  hierarchy of indivisible parts joined into components, parts and
  components put together as assemblies, assemblies connected
  or linked to form systems, and systems integrated into complex
  products doesn't seem clear, and it is unlikely to be all-inclusive.

- The chain of cause–mode–effect isn't clear. Modes are confused
  with effects and sometimes with causes. If this situation occurs,
  the study is virtually worthless. We'll discuss this in much greater
  detail in Chapters 4 through 8.

- The first few pages of the DFMEA seem to be intensive and even
  useful. However, as you read more deeply into the study, the level
  of insight and constructive value declines, often steeply.

- The controls are heavily tilted toward detection, with particular
  emphasis on physical testing of prototypes. Moreover, design
  verification activities seem to have little or no direct relationship to
  the DFMEA study. While this can occur without a form mentality,
  it is almost always part of the "just fill out the form" mind-set.

- Old DFMEA results are used to create new results. Sometimes this
  is as simple as just changing the part numbers and other identifying
  information on the study. In some cases, you will see portions of
  the study that don't even seem to match the design that's under
  consideration.[5]

- There are few if any meaningful insights in the DFMEA study;
  instead, the results are presented as "all is well," an idea that might
  be true but is more likely disastrously false.

---

5. While creating DFMEA "templates" for families of design is useful, a "form
   first" approach usually results in superficial or misleading results, with design
   weaknesses from previous generations carried over to new designs.

When I was first required to carry out DFMEA studies in the 1970s, we used to sarcastically say that the most important thing we needed in order to do a study was a bottle of Wite-Out or Liquid Paper so that we could change the part numbers on a copy of an old DFMEA.

Paper, a copy machine, an IBM Selectric typewriter, and a bottle of Liquid Paper—that's what made DFMEA studies go.

Getting the form filled out and properly filed away was paramount. DFMEA was seen as an irritant, an activity that was used to "cover your ass." It was seen as time-consuming and not really an engineering activity—it was something the quality guys needed for some reason or another.

Of course, there was almost no training available for DFMEA in those days, and the training that did exist was, at best, shallow. In fact, many of the training sessions in this era were focused on showing people how to complete the forms rather than concentrating on the underlying techniques and methodology. FMEA seminars would be perhaps two hours long—a time frame that is ridiculously short given the complexity of any type of FMEA study.

The Third Common Problem: Blaming Manufacturing for Everything

One of the most annoying things that I see in DFMEA studies is a systematic shift of responsibility from design engineering to manufacturing operations. When I see controls that say "check at assembly" or causes that say "hole bored incorrectly" in a DFMEA study, I know that the rest of the DFMEA is suspect and quite probably not worth the paper it's written on.

While this can be a symptom of a much larger problem—such as overstuffed egos in the design group, or open warfare between design and manufacturing engineering departments—this can be easily overcome by using good technique. We'll see how you can prevent this in Chapters 4 and 5. This won't solve deep-seated organizational problems, but good technique will help in making the control of risk a more rational and deductive issue.

The Fourth Common Problem: After-the-Fact Detection

Finding problems in designs late in the product development cycle can be a disaster. The Airbus A380 program, which continues to fall behind in meeting promised delivery dates, cost EADS, Airbus's parent, as much as $6 billion in lost earnings as of early 2009—and the meter keeps running.

What happened on this program? According to the executive, Christian Streiff, who was hired in 2006 to fix things at Airbus when the A380 project started to go very badly, there were design errors in the wiring harnesses for the aircraft. The wiring on a commercial jetliner can be extremely complex; not only is the airplane controlled by electric signals ("fly by wire"), but there are lights and entertainment modules—and more recently, AC power ports—at every seat. The A380, the largest commercial jetliner ever conceived, has more than 300 miles—or 500 kilometers—of wiring.

What happened at Airbus was that two different groups, working in different locations, used different versions of computer-aided design (CAD) software to design portions of the harness system. One group was working on the front half of the airplane, where the pilot's control interfaces and central processors are located; the other group worked on the rear, which includes the control surface actuators, galleys, lavatories, and passenger seating. When the first harnesses arrived and workers attempted to install them into the A380, they found that portions of the harnesses couldn't be routed through the aircraft and that some connectors were incompatible.

This is the kind of thing that can easily be identified in a system FMEA, but either no system FMEA was performed (which I think is likely, given the information Airbus has released about this snafu), or it was done very poorly.[6]

Of course, not everyone works on something as complicated as the A380. But the same thing happens almost every day in most companies that design products. And it happens even when DFMEA is part of the development system.

What causes this is simple: most design engineers, in the deepest recesses of their hearts, truly believe that they won't make errors when they churn out a design. Software engineers may be an exception—they expect that they will make errors—but they erroneously assume that if they test the software enough, they'll find all of the bugs and be able to fix them.

Good design practices used by motivated and clever engineers can yield a product that functions properly. Great design not only yields products that perform well, but does so with a minimum of problems—and with ideas and features that attract or even excite customers. If DFMEA is used correctly, it is, in my long experience, the single most powerful (and logical) set of activities that can elevate a good design to a great design.

---

6. The underlying organizational issues at Airbus—with German, French, and other European groups all working on the airplane at different sites—were a fundamental factor driving this mess. Nevertheless, a broad SFMEA study would have easily identified the risk of front-to-rear wiring difficulties. That's not to say that FMEA would solve the organizational dynamics problems at Airbus, but at least the issue would be clear to everyone.

A variation on this theme—a practice that is often related to or even part of the "form first" mentality—is conducting the DFMEA study after a design is released. After all, once all the testing, debugging, and development is done, it should be possible to provide a better assessment of risk, right?

Actually, this is one of the top ten crazy things that you can do with DFMEA. Anyone who does this will probably confirm that the final design is great and that all risks that can possibly be addressed have already been corrected. There will be a good deal of effort required to prove this, but it will cover your behind, of course.

This is one of the easiest things to see through if a major problem develops after release of the design. When you look for the issue that drives the problem, you either won't find it or you'll find unreasonably optimistic assessments of the underlying cause–mode–effect chain. A DFMEA study performed in this manner has accomplished nothing except consuming scarce resources and creating a false and undeserved sense of security for management.

In summary, there are several approaches to DFMEA that aren't particularly useful or powerful. The fact that these practices are fairly common accounts for the reason that most people who work on DFMEA studies hate them. They can see that a DFMEA done with one of these "bad" techniques will be time-consuming and not very valuable—in other words, a waste of time.

How can you turn DFMEA into the powerful tool that it can become when done well? How should DFMEA be approached? We'll tackle that in the next chapter.

2

The Right Way to Use DFMEA

I f you really want to improve product designs, you must do more than conceive and develop ideas using intuitive and inductive thinking. While innovation and creativity—which are driven by insight and inductive generalizations—are critically important in today's competitive world, inspired ideas that are not executed with exquisite attention to detail are, more often than not, doomed to the scrap heap of history.

Inductive thinking, which takes what is known and projects it into areas that are either poorly understood or even unknown, is absolutely critical to creating a great design. As we all know, if you continue to do what you've always done, you'll get what you've always gotten in the past. And that just won't be good enough in any modern competitive environment.

But that doesn't mean deductive logic can't be applied to design. Not only is deductive logic important in design work, it can support inductive and innovative thinking in a way that is stunningly potent. Done well, it can even drive additional inductive, intuitive, and innovative ideas.

To achieve this, though, you must have a process—you must follow a systematic approach or methodology. All processes have the same fundamental structure: an input is converted, through one or more discrete or continuous actions, into something different, something presumably more valuable, which is an output.

The fundamental principle that makes a process useful is simple: if the inputs and process steps, or transformational activities, are consistent, the outputs will be predictable. That doesn't mean that the outputs will be necessarily good or bad, only that they will be consistent.

If the inputs are suitable and the transformation activities are well planned, then the output may, in fact, be suitable, or even outrageously great. Conversely, bad inputs or badly planned process steps that are carried out consistently may lead to unacceptable outputs—but at least these bad outputs would be consistent and predictable.

On the other hand, if inputs are inconsistent—or if the process steps aren't completed in a repeatable and reproducible fashion—the outputs will be inconsistent and unpredictable. At times, the results may be good, or even spectacular. But, given the complexity of design engineering, it's much more likely that a poor outcome will result whenever inputs and process activities are not consistent.

All of this does *not* mean that you can design a world-class product using a "paint by number" approach. You can't. However, you can use structured techniques to understand the positive as well as negative—and often risky—aspects of a design. That requires a set of reproducible steps, and the DFMEA process is a sound way to do this.

To carry out a DFMEA study in a way that is practical and useful, you have to use a consistent and repeatable process. Moreover, that process must be conceived properly so that the results will be useful and constructive. Unfortunately, many of the "common sense" ideas about doing this are counterproductive. More importantly, when left alone, different design teams (or individuals) will cultivate their own way of completing a DFMEA study.

Whenever this occurs in an organization, the results of the study will be unpredictable, and the likelihood is that the DFMEA results won't be worth much. Even with a clever or exciting design, a poorly developed DFMEA means that there will likely be serious problems with the design, either during the development cycle or after customers begin to use the product, or both.

THE BASIC METHODOLOGY: INPUTS

Most people start a DFMEA study by looking at a product and asking, "What might go wrong?" If there's one key error that's made in DFMEA, that's it.

Think about the obtuse logic this question presents. How you can speculate about what might go wrong before you have defined what is supposed to go right? You can do this, but the outcome will be completely inductive, an imagination-based exercise. The possibility that such an assessment will be either comprehensive or meticulous is very low, probably approaching zero. Some issues of excellence will be omitted and others will lack depth, which means that problems are likely to slip through to the marketplace.

And yet that's exactly how most people start a DFMEA study—they start by brainstorming failure modes.

Further, if you start with the "what could go wrong" question, that's simply not process-oriented because that doesn't really use a structured set of inputs. So, we need to ask a simple question: what are the inputs to a good DFMEA study?

The first critical input is a properly constituted team. If one person completes the DFMEA study, it's generally a futile exercise. This occurs for several reasons:

- A single viewpoint nearly always misses a few important issues.

- If the product is at all complicated, the DFMEA will be fairly long; it might be dozens or even hundreds of pages before all is said and done. If an individual completes the study, it's extremely unlikely that that one person will have knowledge of all of the issues—and it's also likely that she or he will not have sufficient time or energy to do the complete study well.

- No one will ever read or understand the study except the person who wrote it, particularly if the product is complex.

If you are going to work on a DFMEA study as a group, who should participate in the study? How should they work together?

To start with, DFMEA can not be done in a series of one-hour meetings or by e-mail exchanges. In my experience, a team needs to block out three or four hours at a time, turn off the mobile phones, and limit access to e-mail. The work is so intense that dividing your attention with multitasking is nearly always negative and can even be ruinous.

If the team is in different geographical locations, conference calls are a minimum requirement; teleconferences or net-meeting techniques are better, and, if the project is complex enough or has a critical business aspect, traveling to work together face to face may be essential.[1]

Team size is also critical. If you don't have at least three people working on the study, it really doesn't qualify as a team. But if you have more than six people on a DFMEA team, it can easily become unwieldy and impractical.[2]

---

1. I have seen some use of high definition (HD) teleconferencing for team meetings; this is nearly as good as "being there" *as long as everyone involved has at least been introduced on a live, face-to-face basis.* Even if the introductions haven't been made in previous projects, HD conferencing is very good as long as everyone obeys reasonable protocol—minimum muting, not hiding from the camera, and so forth.

2. There are some exceptions to this when complex projects are properly dissected or scoped. We'll come back to this in Chapter 3.

If you want to avoid a narrow viewpoint, you need a wider view. In DFMEA, this means someone who knows what the customer wants, and this may be a technical marketing or sales representative. This is particularly true if the lead design engineer hasn't had a great deal of contact with customers on the product being considered before starting DFMEA activities.

And contact means more than just "I think I know what the market wants." Too often, I see engineers who approach customer orientation in a completely self-centered way. Most engineers get up each morning, look in the mirror, and think, "I'm a normal person." Almost no one thinks they're odd or unusual. Then, this belief in normalcy leads to another conclusion, namely "Since I'm normal, almost everyone else will like what I like; therefore, I will design something that I think is good and everyone else will see it the same way."

Unfortunately (at least, most engineers think it's unfortunate), the worldwide demographic of engineers is probably less than 1% of most customer baselines. As a result, too many engineers design for other engineers, and the results are often poor or even confusing. BMW's initial "iDrive" vehicle control system, which used a very logical but extraordinarily complex system of menus and a mouse-joystick device—a system that many engineers would probably see as "elegant"—is typical of what can occur. The first-generation iDrive, introduced on the 7 Series luxury sedans, was so confusing and difficult to use for most non-engineers that some new owners returned the car within a week of purchase and demanded a full refund.

You also need someone who understands your organization's product technology and history, someone who understands the manufacturing issues that will affect the product, and someone who is savvy about the quality system in your business. If subcontractors are expected to design significant elements of your project, you may need their direct participation as well. Alternatively, a well-versed buyer or purchasing agent may represent a supplier's views.

All of these roles usually lead to a core team, consisting of a design engineer (possibly supplemented with a designer or "tube jockey" who does most of the heavy computer-aided design work), a manufacturing or process engineer, and a quality specialist. These people need to participate in nearly all of the meetings that will be needed to complete a sound DFMEA study.

If the product also includes software, a software engineer may also be necessary—even when the mechanical and/or electrical elements of the system are being analyzed. After all, if the mechanical and electrical elements of a design aren't correct, software can rarely correct this. Moreover, the peculiarities of the physical design elements will have an important impact on the software and vice versa.[3]

Why is the manufacturing engineer important? To begin with, once a sound DFMEA study is completed, much of the design will be firmly entrenched. That means if there are any features that are difficult to fabricate or assemble, or any other troubling manufacturing issues, these need to be identified during the early stages of the product development process. Once the design is strongly established, changes to facilitate manufacturing are difficult, expensive, and—in most cases—nearly impossible to implement.[4]

That doesn't mean that DFMEA is a particularly good tool for identifying manufacturing or assembly challenges, but it is almost impossible to get a design change that will aid manufacturability or simplify assembly once the design has passed through the first stages of verification. As we'll see in a moment, DFMEA is a central element in the design verification process, and, after the first iteration of the DFMEA is complete, it becomes progressively more difficult to change the design for any reason other than substandard performance.

The second critical input is a concept design. What do I mean by this?

A concept design is a design that has sufficient detail to permit a rough, first-pass cost analysis. It should not include detailed or firm decisions about tolerances, and may include only general material selection information. In a mechanical system, this may consist of a layout drawing with the individual components and major dimensions identified but not yet drawn in detail. In an electrical system, this may be nothing more than a schematic or a printed circuit table. For software, a concept design may consist of a logic or ladder diagram. A bill of materials and a draft manufacturing outline are also part of a concept design; both of these items are needed to develop a rough cost estimate.

To the greatest extent possible, a concept design should *not* be worked out in full detail. There are two issues that will hinder a DFMEA study if a fully developed design is used:

- Once design engineers have developed a design in detail and made major decisions about tolerances, features, components, materials, and possibly manufacturing processes, it will be more expensive

---

3. We'll discuss software engineering in a bit more detail in Chapter 3; it is important to note that the principles of DFMEA can be applied to software quite well.

4. Best practice would be to conduct design for manufacturability and design for assembly studies (DFM/DFA) in parallel with the DFMEA studies.

and time-consuming to make any changes that the DFMEA may suggest are needed. If a design has reached a "firm cost" stage, the total effort to change anything becomes quite significant.

- Worse, once a design engineer has fully explored a drawing and spent considerable effort in doing so, most engineers have fallen in love with their design. No matter what they might claim, they will have great difficulty in seeing any weaknesses or flaws in the design. As a result, they will probably—though subconsciously—skew the DFMEA process to prove that their judgment has, in fact, been sound. After all, if an engineer thought some aspect of a design was less than stellar, he or she would never have included that feature, tolerance, specification, or material selection in the design.

Finally, a concept design can and perhaps should be derived from other analytical techniques, such as quality function deployment, structured innovation techniques, and the strategic market plans of your organization. No matter where it came from, or how it was developed, a concept design is still one of the basic inputs to the DFMEA process.

The overall constraints for a concept design must include assumptions about how the product will be used, what types of misuse might occur, and what the cost of the product must be in order to be commercially viable. We'll discuss this a bit more in the section on design verification later in this chapter, but a full discussion of how to set general but comprehensive design goals is a bit beyond the scope of this book. But if the design goals aren't well conceived, the best DFMEA can not prevent problems in the marketplace.

The third and last input for a good DFMEA study is knowledge of the DFMEA process itself. If you read and follow the guidelines in this book, you will have that knowledge. On the other hand, if you approach DFMEA studies with inconsistent, weak, or variable methods, I can honestly say that it's extremely unlikely that the analysis will be worth the effort.

THE BASIC METHODOLOGY: THE PROCESS

Once you have assembled the inputs, you are ready to start, and if you follow the proper steps, you'll probably be able to get a DFMEA that offers you some value. Here is an outline of these steps:

- Define the scope of your study. We'll look at detailed methods to do this in Chapter 3.

- Understand the function of the product envisioned in the concept design. This can be difficult, but we'll see how this can be done in Chapter 4. The key issue in function analysis is to be thorough, reasonably comprehensive, and quantitative if at all possible.

- With a comprehensive and detailed understanding of function, you are ready to assess potential failure modes. We'll work through this in Chapter 5, and if you've done FMEA work previously, you may be surprised at how straightforward this can be when using a deductive methodology.

- Next, you need to determine what can result when something goes wrong and how bad this might be. That's the domain of "effects and severity," and we'll see how to use deductive methods to understand this in Chapter 6.

- Once you know what is supposed to occur (function), what can go wrong (mode), and what the consequences might be (effects and severity), you are ready to look for potential causal factors. In Chapter 7, we'll see how logical techniques can lead to a constructive view of causes.

- In Chapter 8, we'll look at the issue of controls. We'll address controls in a general way in a moment, but getting to a reasonably comprehensive list of controls is one of the key objectives in any DFMEA study.

- Last but not least, we'll take a relatively deep dive into the issue of risk in Chapter 9.

Since most simple processes follow a linear flow—and DFMEA is a linear process—it might be easier to visualize this process with a process flow diagram. Figure 2.1 shows what a deductive DFMEA process looks like as a flow diagram.

THE BASIC METHODOLOGY: OUTPUTS

If you want consistent and worthwhile results, you must use a sound process with a well-defined list of outputs. The deductive DFMEA process is no different. There is a very general (and quite important) overall output, which is the reduction of risk and improvement of customer satisfaction—if there's a *raison d'être* for DFMEA, that's it.

To get to that point, though, there are a number of intermediate outputs, which we'll look at in the following chapters. However, the most important

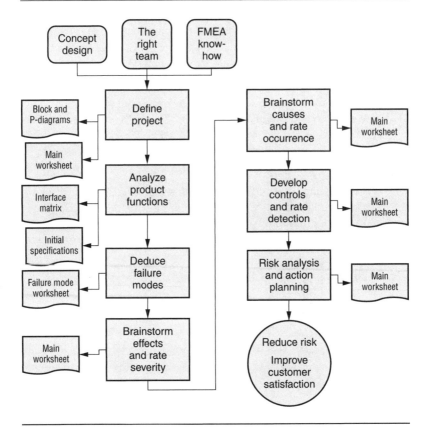

Figure 2.1   Deductive DFMEA process flow diagram.

intermediate output—some might view it as the most critical output—is the main worksheet. Despite the warnings I've given in Chapter 1 about a form mentality, the main DFMEA form is, in fact, very important.

I would strongly suggest, though, that the form is only a means to an end. And that end is a better product with well-defined risk.

While we'll explore the specific issues and criteria that are included in a comprehensive set of DFMEA process outputs in the rest of the book, it is critical for everyone—but particularly those who manage and direct the design process in an organization—to understand the basic theme of DFMEA and be able to provide a meaningful and probing review of the results.

That theme—and overriding goal—is simple: to eliminate errors from the design before the design is released for production development.

However, as I explained in Chapter 1, the best way to do this is to emphasize prevention rather than detection. This specifically means shifting from a reliance on physical testing to more trust in analytical assessment of the design. Physical testing isn't obsolete, but it is far less effective than most engineers believe. The development of the airplane and the contribution of Orville and Wilbur Wright to engineering in general illustrate the power of this idea extremely well.

THE STORY OF THE WRIGHT BROTHERS

Before the Centennial of Flight celebration in 2003, I hadn't known much about the Wright brothers and their efforts to develop powered flight. I suppose I had a fairly common picture in my mind about Orville and Wilbur. Two guys with funny names from a little town in Ohio, who were bicycle mechanics and lacked formal schooling, and had used trial-and-error methods to accomplish something that looked inevitable.

As I watched and read about the Wrights' actual work in 2003, I came to realize how terribly wrong I was. While the Wrights did a great deal of experimental work, most of their efforts were aimed at calculation and analysis. Their triumph did not occur because they were lucky or because all of the technology they needed was present and they just needed to figure out how to put it together—which is what Henry Ford did at about the same time. Ford didn't invent much of anything, but the Wrights were geniuses who have never gotten the credit they deserve for what is truly one of the most significant engineering accomplishments in human history.

Instead, Orville and Wilbur—largely home schooled and without much mathematical skill beyond algebra, geometry, and trigonometry—worked out all of the issues needed to create an airplane. And, most impressively, they used a combination of inductive and deductive thinking to achieve this. In the early stages of their work, their efforts were—like leading-edge product research—dominated by inductive thinking. As they moved toward a serious concept design, their work became more and more deductive and less driven by imagination. In short, they were applying the logic of analytical prevention of design flaws, as opposed to a complete reliance on trial-and-error for detection of design failings.

To begin with, the Wrights were very successful businessmen; their bicycle business in Dayton, Ohio, was a cash machine. Their success gave them the capability to throw almost all of their time and energy into the "airplane project." And, while most other inventors and "aeronauts" (as they were called) were experimenting with gliders and mounting powerful engines on gliders, the Wrights were more interested in the fundamental principles that would govern flight.

In particular, they were absorbed with the pioneering aeronautical research of Otto Lilienthal, a German physicist and one of the most successful aeronauts in the world. In many circles, Lilienthal had begun to be regarded as the greatest physicist since Isaac Newton, and his pioneering work on aerodynamics still provides the fundamental physical laws that govern all flight.

However, Lilienthal is, except in aerospace circles and his native Germany, largely unknown today. Why? The answer is quite simple: like nearly all of the early pioneers of flight, Lilienthal was killed in 1895 when his glider crashed. His failure has, to a large extent, diminished his place in history. And it seems some of his calculations were flawed, which probably contributed significantly to his death.

As the Wrights started their project, they realized (by using Lilienthal's research) they would need strong winds to increase airflow over the wings of an airplane and, through a detailed research process, selected Kitty Hawk and Kill Devil Hills in North Carolina as the place they would use for development. However, they also realized that they would only be able to spend a few weeks a year at Kitty Hawk, when the winds were favorable in autumn. What did they do with the rest of their time?

By and large, they applied analytical methods to design their machines. Using the equations of flight that Lilienthal had published and adding some of Wilbur's unique insights about the application of these equations, they built scale model gliders and used these to evaluate different concepts. Eventually, the work yielded a large glider that could be piloted—an early concept design or "R&D" model—that they transported to Kitty Hawk to evaluate.[5]

When they flew this glider in 1900, they were dismayed to find that it didn't behave the way Lilienthal's equations predicted. In any airplane, flight is possible when lift exceeds drag, and what they discovered was that the lift developed from their calculated wing shape and area was only about one-third of that predicted by the equations.

They returned to Dayton and began additional fundamental research—not design work—to discover what was wrong. First, they devised a simple wheel mechanism that they mounted horizontally on the handlebars of a bicycle. Pedaling as fast as they could go, they watched the rotation of

---

5. Moving their base of operations temporarily from Dayton to North Carolina was no mean feat in those days; it required significant logistical planning and money. However, the Wrights were expending only a fraction of the money that other researchers, like Eugene Langley, were spending.

the wheel when various airfoils were mounted perpendicular to the free-spinning horizontal wheel. From this experiment, they discovered that the drag coefficient they were using was incorrect.

Then, they set out to determine the fundamental properties of air pressure and airfoils—and they invented the wind tunnel to do this. How many people know that the Wright brothers invented the wind tunnel?

Using incredibly clever devices made from hacksaw blades, wire, and wheel spokes, they measured lift and drag in their wind tunnel for more than 200 airfoils and achieved a surprisingly high level of accuracy. When they took these values and applied them to the equations that Lilienthal had published, the results were in very close agreement with what they had observed in 1900 at Kitty Hawk, when their glider wouldn't perform the way they thought it would.

While there were many difficulties yet to be resolved, the Wrights never again had difficulty with the basic issues of lift and drag. They applied this research finding in their future work and, again blending induction and deduction, moved toward the goal of powered flight.

They built and flew more gliders at Kitty Hawk in 1901 and 1902 with the aim of developing an integrated model of flight control, but, by and large, this was a creative—or inductive—application of the deductive lift and drag equations that they had wrestled with in late 1900 and early 1901.

After the 1902 glider flights, the Wrights felt they had largely tamed the issues of lift, drag, and three-axis control and filed a patent for that combination of ideas. They then turned their attention to powering the aircraft.

There were two aspects of this problem that were taxing—devising a practical propeller and obtaining a suitable power source for the system.

The propeller was very challenging. Up to that point, most experimenters had been using two different approaches. One was a flat paddle on a stick, like a Dutch windmill blade. The other was a marine propeller—the type of device used to move a ship through the ocean. There was no published research on either of these approaches, but the Wrights soon discovered that the efficiency of either design was impossibly inadequate.

Neither type of propeller would convert enough engine power into aerodynamic thrust to be practical. The rate of conversion was below 30% for the marine design, and the windmill design was hopeless. As a result, a huge engine would be needed to push the airplane through the air with sufficient speed to create enough lift to fly. However, the engine would be so heavy that the thrust required to lift this weight would be even greater.

In essence, the dynamic trade-off between weight, thrust, and lift was unattainable—yet another analytical finding. So, Wilbur, in what may have been one of the great intuitive leaps in the history of engineering, asked a simple question: what is a propeller? His answer, well known to beginning aerospace students, was, at the time, a blinding insight: it's just a wing that's rotating in a vertical plane—and the resulting rearward, horizontal lift becomes thrust.

With this insight, the Wrights then realized that lift would change at different points along the propeller because the air speed would be close to zero near the propeller hub and highest at the outer tip. This caused them to first calculate and then design a continuous twist in the shape of the propeller; this deformed wing enabled them to get a huge increase in lift, which in turn translated into thrust.

Wilbur Wright estimated the efficiency of the initial propeller at 66%, which was more than double the best result ever achieved at that time. However, in 2003 this design was recreated using the original Wright drawings, materials, and woodworking tools. When this replica was subjected to wind tunnel testing, efficiency was at least 75% and reached a peak of 82%.

This is really a staggering result because the best commercial propellers available today are only 86% efficient. In other words, the Wrights achieved a near-perfect result with their very first design—*simply by applying analytical tools*. The first propeller they built was the one they used in their 1903 Flyer—the first craft to achieve powered flight.

When I learned this story, I started to think that Wilbur might have been the greatest engineer in human history. However, there's even more to think about.

Once the propeller design was understood, the Wrights then calculated the power-to-weight ratio their engine would need to have. They sent out inquiries to engine manufacturers and they were told their requirement was impossible.

So, they decided to build their own engine. Working with their head bicycle shop mechanic Charlie Taylor, they designed and built a lightweight engine. This engine used primitive fuel injection and was made from aluminum, a material that had become commercially available at just that moment in time. (The aluminum was purchased from the firm that would become Alcoa a few years later.)

They completed the engine in less than six weeks. After a few initial failures, the engine ran without incident—and exceeded the calculated level of power that the Wrights knew was necessary from their analyses. This engine was also re-created in 2003 for the Centennial of Flight, and the details of this design are really quite amazing.

Again, calculation was the basis for the fundamental design. And, when all of the pieces were put together—lift and drag, propeller and engine, three axes of control—the result was the first powered flights in 1903.

How revolutionary was this result? Over the next five years, the Wrights refined their concept design and by 1908 had an airplane that could take off, land, turn, and even stay aloft for an hour. After news of their initial flight had leaked out, other experimenters began to fly, too—but all they could do was rise up into the air for a few seconds and return to the earth.

Only the Wrights had developed a practical system of flight control combined with lift to make a practical airplane. When they finally displayed their result in 1908 in Paris, the world was stunned. Soon, other aviators copied the Wrights' work and, in just a few years, the competitive pressure of World War I pushed aeronautical technology to heights that could not have been imagined just 15 years earlier. In retrospect, it appears that the Wrights made a 25-year leap in technology in less than a decade, providing all of the fundamental expertise that remains at the heart of every airplane flying today.

And they did it with analytical tools more than with experimentation. While they did engage in trial-and-error testing, it was more as an adjunct to their analytical work. They derived fundamental principles and measured important properties of airflow with experiments, but most of their work was calculation-driven—and not primarily test-driven. They tested to confirm their analyses, but they were determined to avoid the fate that had befallen Lilienthal and so many others.

In many ways, they applied the ideas that drive sound usage of DFMEA—and Charlie Taylor said as much in a 1948 interview when he observed, "Those two sure knew their physics. I guess that's why they always knew what they were doing and hardly ever guessed at anything."

To put this in perspective, we could say that the Wright brothers effectively invented analytical design, which is the foundation on which prevention-based development is built. Physical testing is not and probably never will be obsolete. However, *relying on physical testing as the primary or even sole tool of development is a prescription for disaster*—or, in some cases, death.

THE CHAIN OF VERIFICATION: FMEA AND DESIGN SYSTEMS

As usual, let's begin by defining more terms. Every design really ought to be verified before significant manufacturing development begins. What does it mean to undertake design verification—or "DV" as it is usually referred to?

Wikipedia offers this definition: "the act of reviewing, inspecting, or testing, in order to establish and document that a product, service, or system meets regulatory or technical standards."

A more powerful definition for design verification can be stated as:

> Verification is the process of seeking and generating information, either in the form of analysis or physical test results, that provides sound evidence that a given design meets or exceeds all known and reasonably anticipated customer or market requirements for a product.

Let's dissect this a bit and see what implications this approach has for design. First, the goal is to gather and/or develop realistic proof—not absolute proof—that a design does what it's supposed to do. Second, the bar for verification is set higher than just regulatory or technical standards; it includes all reasonable and anticipated customer usage factors. However, it can and almost certainly should exclude the kind of deliberate abuse that is common for some products. Even within those criteria, it might be that the product has to preserve safety or continue safe operation in the event of some types of customer misuse.

For example, should an automobile be designed to withstand a crash? Sure, but what kind of crash is reasonable? If a small car were designed to withstand a head-to-head crash at Autobahn speeds, how much would it cost? Is that a practical and reasonable customer usage factor? There are no absolute answers in any case, but determining the overall market goals and technical goals for a product or system is a major management responsibility.[6]

Third, a product design can be verified with a combination of analysis and physical testing. However, design verification does *not* require that actual commercial manufacturing (or service process delivery) be included in the assessment; the inclusion of actual manufacturing is the realm of validation—which is not the same as verification.[7]

---

6. Quality function deployment is a good tool for deriving a set of verification objectives or goals and then making sure that the requirements are stated in measurable technical terms. Nevertheless, understanding the market requirements for a design remains a critical DFMEA process input, one that can not be generated by DFMEA in and of itself. As an aside, this issue is also quite relevant to the Pinto saga described in Chapter 1.

7. That's also the domain of PFMEA rather than DFMEA. Verification is about design and uses DFMEA as a tool to determine what's needed for verification, while validation is about production—which uses PFMEA to determine not only what's needed for validation, but for ongoing process control.

To apply analytical design to achieve a sound and rational design verification outcome, you need to follow—wait for it—a structured process. So, how do most engineering groups develop a new product?

As soon as a formal project is approved for development, the initial design work is completed at the earliest opportunity. Just get it done, don't spend much time doing detailed calculations and worrying about problems; these things can be worked out later. This allows prototypes to be constructed as early in the program as possible and testing to be started at the earliest possible date.

Of course, these early prototypes are never successful, so the early tests result in failure. Then, the design is refined, more prototypes are made, and, in nineteen out of twenty cases, the second set of prototypes fails. Then, more design, more testing—and this cycle is usually broken by the need to release the design to meet a commercially determined timing date. The resulting verification evidence is rarely sound, and problems generally persist throughout the commercial life of the product. As a process, this usually follows a sequence that appears much like Figure 2.2.

This approach, which is a classic trial-and-error method, leads to many difficulties.

To begin with, this approach makes it nearly impossible to contain timing and cost plans. Timing is repeatedly disrupted as the design is constantly adjusted based on the most recent test results. The difficulty of managing the variety and complexity of various levels of design (and prototype hardware) can be mind-boggling. And, the need to repeatedly build new prototype samples for testing is costly. Worst of all, there's no real reason to believe that any promised completion date will, in fact, be met.

Ultimately, testing almost never reveals all of the flaws in the design. The statistical significance of most test programs is terrible, and testing

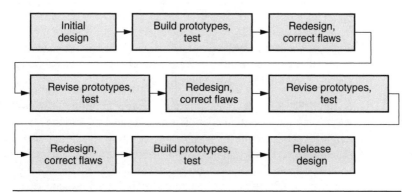

Figure 2.2 Typical product development process.

rarely duplicates the exact conditions that will be encountered in the marketplace.

Passing a test does not demonstrate that a design will do what it is supposed to do. Failing a test could (but doesn't necessarily) reveal design flaws, but passing a test only demonstrates that the particular prototype sample subjected to testing performed acceptably under the conditions of the specific test that was actually carried out.

The weaknesses in test-dominated verification information are substantial and many. Here are a few of the most common:

- Tests rarely reproduce the actual usage conditions encountered in the marketplace. Subtle differences between lab-based test conditions and customer usage are inherent in controlled testing.

- Successful testing has a very low level of statistical significance. This is driven by many issues, including:

  - Prototype test articles are virtually never produced at the limits of tolerance, whether those limits are physical—such as mechanical size tolerances, electrical properties (resistance, capacitance, and inductance)—or consist of material properties, such as strength, elasticity, fatigue resistance, conductivity, and so forth. (I'd love to be listening to the discussion when a steel sample is ordered from a mill at "minimum strength" for a prototype; if a steel mill knew how to control strength with that kind of precision, the world would be a far different place.)

  - Most real-world marketplace problems are relatively rare. If we assume that all components in an assembly were made according to the released drawing—and then assembled in accordance with all plans and specifications—some articles would very likely *still* fail. This would occur because the design weakness might not be recognized unless all key parts and components were fabricated and assembled at a stack-up of minimum conditions—a state of affairs that we've already admitted is impossible to achieve in nearly all real-world product development programs.

  - Durability testing that does not extend testing to failure provides very limited information about failure rates and underlying reliability. However, test lab managers hate test-to-failure schedules—as do project managers—because the time (and cost) to complete such testing is indeterminate. Similarly, testing one piece to failure (a common alternative) doesn't provide much information, either. Unless dozens of samples are tested to a limit that is short of failure, the statistical significance of the results is limited and, for safety-related products, frighteningly inadequate.

- Engineers have an amazing ability to rationalize unsuccessful test results. Too often, prototype articles fail in a test and the results are then discarded, discounted, or overlooked. We'll explore that in a minute in the story of the Apollo I disaster.

- Field testing is, like testing a few samples for a durability limit, not particularly revealing. How often is field testing representative of the range of marketplace usage? Again, dozens of tests must be conducted; one test is just a point-sample with a 50% confidence interval. It takes dozens of samples to reach a 90% confidence interval and hundreds of samples to reach a 95% confidence interval. In a Six Sigma world, is that sufficient?

- Even well-planned test regimes are rarely comprehensive and almost never address all of the most common failure scenarios for a new product.

- What happens in the test lab stays in the test lab. Beyond the tendency of product engineers to rationalize that unsuccessful results are really okay, lab engineers and technicians have a strong tendency to overlook key nonstandard test issues and elements and to destroy important evidence that might well shed light on fundamental design weaknesses. (I must add that this isn't malicious or, in most cases, deliberate. It often arises because lab operations are too often seen as horribly inefficient cost centers by executives, and the lab staff struggles to make it clear that they are important contributors to the company's success. I wish I had a dime for every time I heard an executive moan about test equipment that sits idle and isn't in use every day, all the time—but that's another story for another day.)

So, what's a better way to do this?

In my experience working with dozens of product development processes around the world, I have concluded that there are at least eight major goals that should be addressed in improving the design verification process in any organization:

1. The process should take a minimum of time and money and have a relatively low level of uncertainty about both issues. In other words, the process should, within limits, be on time and within budget.

2. Verification of the design should be, to the limits of the commercial goals set for the product, comprehensive and not merely based on historical records of what has gone wrong in the past.

3. Once a design is released for detailed manufacturing development, the probability of design changes, *based on design issues and not based on shortcomings in the manufacturing plan,* should be extremely low. In other words, once the design is properly verified, design activities should be largely closed for that product.

4. There should be sufficient early involvement of manufacturing or production personnel in the design process so that obvious and clear difficulties in fabrication, assembly, and transport of materials and control of work-in-process are eliminated before the completion of verification. Manufacturing development should then be centered on manufacturing issues, not design issues.

5. Management should use verification results to understand and decide which risks the business is willing to take and which are not acceptable. However, once risks are accepted, management should *not* be surprised or disappointed that problems related to accepted risk factors arise in the marketplace.

6. The release of significant capital for manufacturing tools, machinery, and equipment should be coupled to design verification results. In the best practice, a design should be fully verified before significant investment for manufacturing is made; this will reduce the probability that capital budgets will be exceeded.

7. To the greatest extent possible, analytical methods should be used to debug designs. Physical tests and demonstrations should largely be confined to confirming analytical methods and results.

8. Funds for prototypes and testing should not be released unless and until all of the analytical elements of verification have been completed and any design weaknesses discovered have been rectified.

Of course, all of this must be done with minimum capital while generating world-class products of unsurpassed quality. That's a tall order, but any company can move significantly in this direction—and DFMEA is the keystone element of doing so.

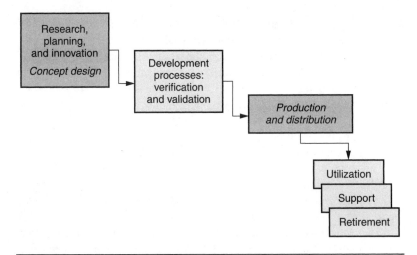

Figure 2.3   Separation of innovation and development processes.

To start, any business needs to understand that "R and D" is almost impossible to carry out under the discipline of a budgeted, planned product release. I've personally witnessed (and helped rectify) several major commercial disasters that were really R&D projects that metastasized into deadline-driven commercial programs.

So, the first thing that executive management must do is keep true research and development work separate from commercial development whenever possible. Schematically, the overall cycle should follow the general steps shown in Figure 2.3.[8]

While innovation or invention is important in product development, major or significant advances in technology, either product or process based, should be confined to the greatest degree possible to R&D operations and not creep into verification, validation, or production and distribution.[9]

This is a huge challenge for some firms—they tend to shun or even deride efforts to conduct long-range product and technology planning—but only minor innovation should be introduced once a concept design has been

---

8. These steps are outlined in ISO 15288; although this standard generally applies to aerospace development, it's applicable, in broad terms, to almost any commercial product.

9. This is yet another subject that could consume a complete book.

created. There's never enough time or money to cram breakthrough developments into a tightly paced development project schedule.

This presents a deeper question about how any design organization releases a concept design for a funded development process. If the concept design is too nebulous, and the underlying attitude is "I don't know what we need for the marketplace, but we'll know what it is when we see it," then no amount of DFMEA work can really fix that kind of an organizational deficit. An on-time and on-budget result will be some kind of minor miracle if the overall program or project objectives aren't sufficiently clear.

However, if the discipline suggested by the simple diagram in Figure 2.3 is in place, then a sound design verification process, or DV process, follows a sequence of events I call "the chain of verification" (see Figure 2.4).

The critical aspect of this process flow is that DFMEA is the centerpiece of the verification process. DFMEA studies are done on concept-level designs, with design for manufacturing/design for assembly (DFM/DFA) studies done in parallel, which insures that manufacturing considerations are included in early design decisions.

The DFMEA process then generates a comprehensive list of *controls* that form the basis for the design verification plan, or DV plan. Then and only then should detailed designs, drawings, and specifications be

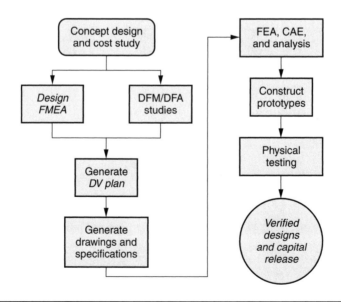

Figure 2.4 The chain of verification.

developed; however, these details are absolutely necessary to carry out finite element analyses (FEA), computer-aided engineering (CAE) studies, simulations, and other review and paper-based study methods.

These intellectual activities, which we will see comprise prevention controls, should be used to find and correct as many flaws as possible before constructing prototypes or doing any testing. If the Wright brothers could do this with paper and pencil, modern design teams—with significant electronic computation and simulation tools—can certainly do this.

Finally, testing should be conducted to confirm what should have been demonstrated with analytical techniques—that the design will, with a high degree of likelihood, perform in the way intended in the initial concept development studies. Again, think about the Wrights. As Charlie Taylor noted, there were hardly any experimental failures during their development work because they had carefully analyzed most of the relevant issues nearly every time.

Done properly, this will turn the looping design–test–redesign–retest–redesign sequence considered earlier into a more direct and predictable sequence of events (see Figure 2.5).

Yes, there may be a test failure or two in even the best-conceived and most thoroughly analyzed design. Wilbur Wright was nearly killed in 1907 in a bad crash outside of Dayton. But the endless loop of design-and-retest should not occur. More importantly, the effort necessary for completion of the design process should occur in a shorter time frame and with less cost—or at least close to budget. Further, manufacturing development and capital expenditures should be better controlled as well.

All of this doesn't mean that a well-managed development and verification process will choke off creativity and lead to mediocre products. It can, but only if the research, planning, and innovation efforts that go into the concept design are pedestrian, tied to safe but middling market plans, and led by managers who think their primary responsibility is to prevent mistakes. Those things suffocate creativity; sound verification practices don't.

The challenge that arises when these ideas are applied is that they require time and resources to conduct the DFMEA and related design

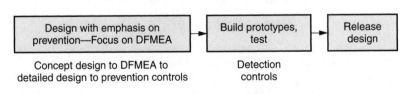

Figure 2.5   Prevention oriented design sequence.

verification planning activities. Too often, executives and senior managers are impatient; project plans are accelerated, and before you know it, you are running down the path of the design–test–redesign–retest loop. Any time saved in the early part of the process is then wasted many times over in the latter stages of the verification process as testing has to be repeated again and again.

Or, it can be much worse. Hurrying through the early stages of verification, particularly DFMEA, can lead to disaster. The story of the Apollo I launchpad fire illustrates this very well.

THE APOLLO I ACCIDENT

Background

The Apollo I capsule, like the older one-man Mercury and two-man Gemini space capsules, was originally designed to have a pure oxygen atmosphere. The rationale behind this design was both historical and practical, even though almost anything that can be oxidized can burn intensely in pure oxygen.

First, pilots had been breathing oxygen in airplanes for decades, so there was plenty of physiological data related to breathing pure oxygen for prolonged periods. The aerospace industry was well versed in the safe utilization of oxygen with intricate equipment—or so they thought. Second, the use of diluting gases—gases that would reduce fire hazard—was complex. Third, the use of pure oxygen reduced system complexity and saved weight, a critical factor in the Apollo program.

In an ideal dilution, oxygen would comprise at least 20% and no more than 40% of the capsule atmosphere, while a second, nontoxic gas would make up the rest. This is very similar to the air we breathe, where 79% of the atmosphere is nitrogen, just a bit more than 20% is oxygen, and the rest is trace gases, including carbon dioxide and monoxide, oxides of nitrogen, and inert gases, primarily neon and argon.

Using nitrogen in a prolonged pressurized condition is risky for humans. It can lead to decompression sickness, or the "bends." It can also cause nitrogen narcosis, or "rapture of the deep," which can seriously impair judgment and reaction time, and can be fatal.

Virtually any gas other than oxygen, including inert gases, will cause narcosis, with the exception of helium. However, using helium in a pressurized breathing environment is also tricky. It causes "Mickey Mouse voice," which anyone who's breathed from a helium balloon has experienced. This could lead to significant communication problems—and could make the space flight experience seem ludicrous.

Helium is also difficult to seal; helium atoms are so small that they tend to squeeze through the tiniest openings and can even diffuse through solid metal. Adding sealing for helium would add weight and complexity—again, negative factors in space travel.

Finally, high concentrations of helium can also cause a sore throat if used for long periods of time, and the Apollo program was going to the moon, so astronauts would have to breathe helium for days on end in some missions.

Even with all of the risks, the capsule designers had recommended a nitrogen–oxygen system for Apollo. NASA managers nixed this, though, citing the success of the Mercury and Gemini programs as evidence that pure oxygen was a manageable risk factor.

Apollo was an extraordinarily high-profile project, and the pressure on NASA to move ahead—or even to beat the declared time goals—was significant. The specter of the Soviet Union landing men on the moon before the United States could do so was a huge political nightmare, and there's no question that NASA felt the weight of this possibility.

This situation probably contributed to this decision as well, although I've not been able to find any direct evidence to support this. Nevertheless, any engineer involved in a high-stakes project knows that the pressure to get things done on time and within budget is always substantial. It would be difficult to believe that this did not color the judgment of management regarding the use of pure oxygen.

The Capsule

The Apollo capsule was much larger and significantly more complex than the Mercury or Gemini capsules. The development process was plagued by many minor problems and flaws, which were identified through laborious testing. The capsule was so problematic during development that the Apollo I Mission Commander, Major Virgil "Gus" Grissom called it a lemon.

The capsule was subjected to a formal, multi-week "design certification review" and was issued a "certificate" on October 7, 1966, stating that the design was flightworthy, pending correction of some open issues.

Testing at the Kennedy Space Center followed, and several failures turned up after the "certificate" was issued. These flaws were both manufacturing-based and design-based. One of the most notable was leakage of water-glycol coolant inside of the capsule; all of these issues were addressed.

Whether these problems were actually resolved, or even fully understood at that point in the program, is a question that has never been and probably never will be answered—again demonstrating how difficult it is to use prototype hardware as a primary tool to debug designs.

After corrective actions for all of the problems were implemented, the capsule was "recertified" by the Apollo Program director in late December of 1966. More tests were then conducted in an altitude chamber, and all tests were concluded successfully. The backup flight crew—not the astronauts who would fly the capsule—"expressed their satisfaction with the condition and performance of the spacecraft" according to the final accident report issued by NASA.

The Accident and the Aftermath

In late January 1967, the capsule was ready for a "plugs out" test, with the capsule mounted on the rocket and the astronauts sealed inside, in which all systems and operational procedures are simulated—with circumstances as close to launch conditions as possible. During this test, a fire erupted in the capsule, which quickly escalated, and within a few seconds the astronauts were asphyxiated, and very likely dead within four minutes.

Their bodies were severely burned, and their nylon spacesuits were partially combusted in the fire. The ground crew couldn't enter the capsule immediately, as there was a very real risk that the escape rockets, which would pull the capsule from the launch rocket in an emergency, might be triggered or even exploded by the fire.

There was evidence that the astronauts attempted to escape, but the release system on the hatch was slow and cumbersome. The hatch opened inward and required one of the astronauts to twist around to release the latching mechanism and pull the hatch in. Finally, because the "plugs out" test was conducted at a slight overpressure of 1.1 bars to simulate orbital conditions—and once the fire started, pressure rose even more—it would have been physically difficult to break the seal and release the door.

Ironically, this design had been adopted in response to another problem that nearly killed Grissom during the Mercury program. When his Liberty Bell 7 capsule splashed down in the Atlantic Ocean, explosive bolts triggered a premature release of the hatch and the capsule sank. Grissom barely escaped drowning and NASA banned explosive hatch release mechanisms.

Investigation of the accident was time-consuming and never reached a definitive conclusion. The investigators believed that a minor malfunction or wire insulation failure, possibly related to glycol-water coolant interaction, led to a spark. In the pure oxygen atmosphere at 1.1 bars, aluminum and nylon will burn intensely—like sap-rich wood—and a flash of fire consumed everything in the cabin. The fire only went out because the oxygen in the capsule and the astronauts' suits was depleted—and then pressure from combustion ruptured the capsule.

The entire sequence of events, from the start of the fire to the rupture of the capsule, took less than twenty seconds. The danger inherent in attempting an immediate rescue made death certain for the astronauts.

After the investigation, dozens of design changes were made, including the replacement of all flammable materials in the spacesuits and cabin. The door was redesigned to open outward and pressurized nitrogen was made available to "blow open" the hatch in an emergency.

More than a thousand wiring changes were made to reduce the risk of sparking. A sixty percent nitrogen atmosphere was introduced at launch, which was then "purged" to pure oxygen after 24 hours to eliminate problems with narcosis in long flights.

Despite extensive and even exhaustive testing, the design was, by any impartial standard, terribly flawed. The risks and cause–error state–effect chains were not well understood, and the analytical design work that had been done was far from comprehensive.

A truly powerful DFMEA study with proper verification of the entire capsule system may well have averted this tragedy. It would have been time-consuming and very expensive. However, it would have consumed far less time and far less money than NASA actually consumed in the aftermath of the tragedy. The delay in the Apollo program was significant, and the budget-busting expense of redesign was perhaps even larger.

3

Step 1—Define the Project

A t this point, you've had a great deal to consider about the way to use DFMEA and the inputs to the fundamental process. We're now ready to tackle some of the real work of DFMEA, and it all starts with proper definition of the project.

From our overall process flow diagram in Figure 2.1, we are now embarking on Step 1 in the DFMEA process. In this step, we'll develop block diagrams, P-diagrams, and start the main worksheet (Figure 3.1).

Creating these diagrams is, without question, a good deal of work. However, it is work that will pay off over the course of any project. In addition—and most importantly—it is work that, in one way or another, eventually gets done. If you don't create these diagrams, you will still face the issues that these diagrams address.

Alternatively, you may (and many design teams often do) miss important issues and design criteria because these tools aren't used. There really is, as Fram reminded everyone for years in oil filter commercials, a "pay me now or pay me later" situation at work here.

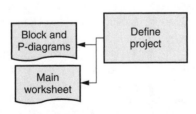

Figure 3.1 Step 1 of DFMEA process.

STARTING OUT—
UNDERSTANDING SCOPE

Another common error in DFMEA processes is poor understanding of the scope of the study. This bewilderment isn't a trivial matter—it is inevitably driven by critical and fundamental decisions made by management about how the design project will be carried out and managed.

However, in most of the cases I have observed in the last twenty years, a sound understanding of the design project scope is rarely achieved. In principle, this is easy. Most engineers would never admit that they don't know the scope of the project they're working on, but in reality the overwhelming majority of projects lack discipline in the way the product is conceived, the work is divided, and overall responsibility for design decision making is parceled out.

In short, it's not the gross level of scope factors that isn't sorted out, it's the fine detail. And in DFMEA, verification, and definitive commercial success, it's the details that matter. Should you doubt that this is a real problem in the twenty-first century, I simply ask you to consider the A380 wiring fiasco described in Chapter 1. How can any executive at Airbus claim that the released design had been adequately verified?

If—and when—you enter into a detailed discussion of scope, you are likely to make several discoveries about the span of any project that are hard to describe but easy to recognize once you've seen them. The first is that a clear breakdown of a product is easy to discuss, but more difficult to carry out. The second is that the way a product is categorized and divided for scope really addresses the overall design philosophy that will be followed during the project. And, how many managers, project leaders, or design engineers have ever realized that every project has an underlying perspective—and that this outlook has a meaningful impact on the project?

The third thing that a sound scope dissection accomplishes is proper modularization of a design. Over the years, I've come to realize that a complex design needs to be broken into interconnected but nevertheless discrete pieces. If any "indivisible" piece is too complex, it can be nearly impossible to sort out the issues, as the interrelationships between design elements become too convoluted to assess and analyze. Inevitably, something can and will get confused or even missed in the fissures or interfaces between design elements.

Part of the difficulty in breaking down a project into a fully understood scope plan is that this is an activity that is partly deductive and partly inductive. There are no right or wrong answers, and the same team could reasonably decide on a different schema of scope for similar projects—or even nearly identical projects at different times.

However, if you fail to understand how the different elements will work together—and how one part of a project might affect other elements—you could be echoing some of the errors that led to the Apollo I disaster. Was the interaction between the wiring, switches, and atmosphere considered carefully? Or was it simply brushed aside without adequate assessment of both the requirements and the consequences?

To understand this issue in a more general sense, let's consider how a clamshell-type mobile phone might be divided into reasonably sized projects.

The first approach could be to attack the project holistically. There are no subdivisions, there's only the phone as a whole. This would likely include several hundred individual elements. If you count the software code needed for a modern cell phone to operate this could include several thousand elements.

How would work be assigned in the design project? No single person could hope to design the entire phone and supporting software; it would be odd to find someone with all of the requisite mechanical, materials, electronic, and software engineering skills needed to do all of the work. Even if this were possible, the sheer volume of work needed to design a moderately complex item like a mobile phone would make it impossible for even the brightest and most experienced engineer to get all of this work done before the concept design had become utterly obsolete.

An alternative way to approach a "holistic" scope would be to have a single project manager with dozens of people reporting directly to the project manager. Then, a really smart engineer might be able to control everything while letting each member of this mega-team have a tightly limited, easy-to-understand scope. In theory, this is the dream that many engineers have of how a project needs to be run. "I am a genius and I can and will control everything. And, when we are done, the product will be better than anything else ever designed because everything was carefully controlled."

This is impractical, too. It's still too complex to handle with a single focal point.[1]

---

1. Steve Jobs may get away with this to some degree at Apple, but even this is risky. Apple's stock price dropped precipitously when Jobs left for several months to deal with serious health problems. And Jobs unquestionably has a level of organizational authority that almost no one else has in any large firm. Further, I'm sure that Jobs simply can't be deeply involved in the design of every subsystem and component on, say, an iPhone.

What else is possible? The first possibility is to divide the project along lines of engineering profession. There would be a mechanical team, an electrical team, and a software team. These teams would have to interact significantly, and the overall specification for the complete phone would need to spell out clearly how the resulting organizational interfaces would be addressed—not to mention the actual performance or technical interfaces.

It wouldn't do to fall into the same trap that Airbus fell into on the A380. However, this is a reasonable division and one that many businesses do follow. It does have a drawback though—professional division tends to cause serious "turf wars" as each professional department wants to see its own area of technology as the dominant and critical piece of the system.

A second division of labor might be along the lines of fabrication and assembly. There might be a design team for the printed circuit board, a design team for software, a design team for component fabrication, and a design team for final assembly.

This, too, would be a reasonable division of labor, although this approach probably has more intricate organizational and technical interfaces than the engineering profession division, particularly if supplier firms are included on the component fabrication team. And again, many companies can and do follow this kind of scheme and, if managed carefully, this line of attack can be a successful way to deal with complex projects.

Finally, the phone could be divided into conceptual elements. A clamshell phone has three major conceptual elements: the upper transceiver, with the screen, camera, and a few of the more esoteric controls; the lower transceiver, with the printed circuit board, battery, and keypad; and the software. Software always tends to be separate in most division-of-labor plans, but we'll consider a way to make software a more integral part of project modularity when we look at block diagrams in more detail.

This is perhaps the most powerful way to divide the design work, because it reflects the actual usage of the phone and is more likely to apply a customer-driven approach to the design. However, it also has multiple organizational and technical interfaces. And that's something we need to recognize: no matter how a multi-element design is broken down, interfaces will be an area of concern.

One caution is necessary at this point, though. There are many organizational factors that enter into a conceptual division of scope scheme—in particular, a deeply functional organization, with a director of mechanical design, a director of electronic design, and a director of software development, will find that traditional lines of authority and senior management prerogatives are disrupted. Working-level engineers may soon develop a greater level of affinity for their project than they will for their "home department" and their functional boss.[2]

In essence, this approach will force a flatter hierarchy in organizations; most observers of industrial organizations assert that this is a positive rather than a negative factor, a conclusion that matches my experience in most businesses. Still, senior managers who prefer to retain the ability to intervene and inject their technical tastes at almost any point in the project may (and often do) rebel against this.

This requires that the senior management group in the business—I'm reluctant to describe such groups as teams, because they rarely behave as a team—must understand and commit to a way of assigning work, assessing performance, and rewarding success that is based on project or programmatic outcomes rather than individual performance. A dazzling bit of software that simply won't support the released hardware design means that neither the software nor the hardware is in any way brilliant.

This further reinforces the general ideas discussed in Chapter 2, where the importance of setting clear program or project goals in the form of a concept design—and then working toward those goals with minimal shifts in direction—is a necessary ingredient in on-time, on-budget design verification.

In any effort to divide a project into manageable pieces, interfaces pose a potentially serious problem. So, whatever is done in DFMEA must include an effort to manage those interfacial issues, including both technical and organizational interactions.

CREATING BLOCK DIAGRAMS

To make sure that interfaces are properly managed, you must have a road map so that everyone can see where the organizational interactions will arise in a complex project. It's even more important, though, to make sure that design issues don't fall through these cracks. And make no mistake: these dividing lines can cause great difficulty if they aren't properly understood and managed. The most powerful way to do this is to create a graphical model or map of these factors.

---

2. This is effectively the opposite of what occurs with an engineering profession division, where these same managers stake out territory and then usually complain about how the *other* groups don't support their work properly.

You can create a map that addresses these issues by sketching out the project in the form of a *block diagram*. A block diagram is simply a logical diagram, somewhat like an organization chart, that records the underlying division of design elements in a project.

Block diagrams typically proceed from the broadest perspective— from the product as sold to customers—to the most detailed perspective, namely that of individual components or software elements. The physical layout of the diagram proceeds from higher levels on the left of the diagram to more-detailed lower levels on the right.[3]

If you were designing a simple one-piece component, like a one-piece brake rotor or perhaps a simple diode, you really wouldn't need a block diagram. However, even a simple trailing arm suspension on the rear of a vehicle can benefit from a block diagram.

Nevertheless, one of the most important things anyone needs to understand is that a complex design will—without any doubt—result in a complex DFMEA.

To grasp what this means, let's take a look at two different examples and see how we can untangle and manage the complexity—or lack of complexity—that may exist in a design project.

A Short Block Diagram Example: The Clamshell Phone

To understand a block diagram, we need to build a simple graphic using nothing but labeled boxes and arrows that show a progressive division, from general to specific, of the elements in a design. Some projects can be captured on a single page, but more elaborate projects can require many pages to see the entire dissection scheme.

Let's start with a clamshell phone,[4] and use the conceptual division of elements that I described earlier. However, let's modify that earlier depiction a bit and place the software in the context of the main circuit board and associated electronic elements. And, I want to emphasize that my goal in this discussion is to illustrate important ideas that arise in creating block

---

3. You can also proceed from top to bottom, but I've found that most diagrams fit more conveniently in left-to-right format; this also means that landscape rather than portrait views of such diagrams are more convenient.

4. This example will be a composite based on several simple, commercial cellular phone sets; no particular phone would likely match the example itself, and that's intentional.

diagrams—not to actually design or describe a tangible or specific mobile phone model.

At the highest level of conceptual design, we can divide this into two major elements—the upper and lower transceiver assemblies. With this approach, the "system" for this project would be the entire phone. The high-level subsystems would include both the upper and lower transceivers, and each of these would have additional subsystem divisions.

We can start by sketching out a relatively simple diagram—anyone could develop a different diagram for the product under consideration—but looking at an example is much clearer than trying to explain a block diagram using words alone. So, let's start with one possible result, shown in Figure 3.2.

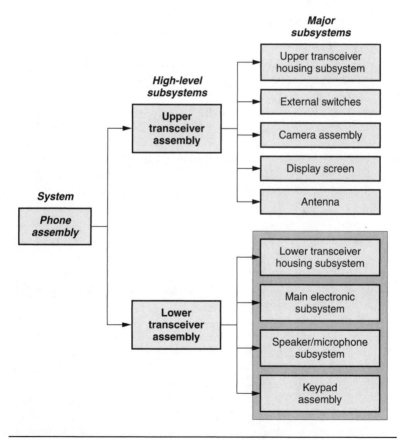

Figure 3.2   Clamshell cellular phone block diagram example.

If you study this diagram, you can see that the structure of the graphic itself is significant. Each "column segment" is a major division in the design—and would likely be mirrored by a similar division of work in project teams.

For example, the highlighted lower transceiver column segment, which includes the housing subsystem, the electronic subsystem, the microphone/speaker subsystem, and the keypad assembly, is very complex. Each one of these subsystems could easily branch into several divisions. The electronic subsystem would likely include the software (which will also be divided into multiple sub-subsystems), the battery, the SIM (subscriber information module) card, the main printed circuit board, and various connectors.

How many different teams would be needed to carry out this work? What skill sets would be needed for each team? What budgets would be necessary? And what level of DFMEA study would be appropriate for each level? What would be the hierarchy of control between the main circuit board team and the software team? Undoubtedly, these are things a phone manufacturer would have a great deal of experience managing, but the specifics are not without risk. Airbus had more than three decades of cross-national experience in managing complex aircraft programs but still failed to manage similar interfaces on the A380.

Moreover, we can see potential interface problems even at the highest level. Which group—the upper or lower transceiver group—would be responsible for the clamshell hinge? Each group would have some aspect of the hinge in their domain, but which project leader would be responsible? What about the routing of wiring from the upper to the lower transceiver?

In addition, as each of the subsystems shown is further divided, the diagram would grow to the right side. New column segments would arise as each subsystem is divided into groups of components, and components are exploded into individual parts. Software would almost certainly be split into several column segments, as would the main circuit board.

As these column segments are developed, the issue of modular design becomes apparent. One of the most important things that should be considered is the number of boxes, or elements, in each column segment. Over the years, I've found that any block diagram with more than about a dozen elements in a column segment indicates a likely problem area. As we'll see in Chapter 4, managing more than about a dozen design interfaces in a single module can create problems because the number of interfaces tends to rise with the square of the number of elements in a module.

So, if you find that a design has seventeen elements in a column segment, you may want to rethink the way the modularity of the design has been approached. Or, if you find that a column segment could have several dozen elements, as it sometimes does in software development, you'll definitely want to revisit the concept itself.[5]

That doesn't mean the problem of complexity goes away. Instead, it is managed on a module-to-module foundation rather than on a component-to-component or element-to-element basis. After long experience with this, I have found that managing complexity at a system level is easier—and clearer—than looking for every interaction between every element.

What that means is that each column segment represents another level of FMEA study that can and perhaps should be carried out. In this example, that probably means several dozen FMEA studies for the phone set, starting with the broadest (and least detailed) studies that will be suggested by the column segment structure of the block diagram.

In the diagram shown in Figure 3.2, there will be three SFMEA studies, including one for the phone system itself in which the focus will be on the interaction between the upper and lower transceivers. This will be the point at which the issues of hinge requirements and electrical connections between the transceivers will be explored and presumably resolved.[6]

There will also be one SFMEA study for each transceiver (one for each column segment). These will be relatively simple studies, perhaps no more than a few pages in length, that can reveal how the major subsystems interact with one another.

As the diagram is further "exploded" to the right, finer levels of detail will emerge. At each new level of detail, new interactions and interfaces will be found and will be dealt with in turn. Finally, as the branching of the exploded diagram reaches the final level of detail—the level of individual parts, components, and software subroutines, DFMEA studies for each column segment become possible.

At this point, I can hear the echoes in many readers' heads:

> That's ridiculous. Completing dozens of SFMEA and DFMEA studies for one project isn't practical. We'd never have the time, money, or people necessary to do this work. There's no way I

---

5. I've actually encountered commercial software that really had no hierarchical structure and no real modularity; in one case, 50,000 lines of code had simply grown like Topsy over time. Whenever another bug was discovered— as customers managed to do on a frequent basis—the entire 50,000 lines of code had to be searched and examined for flaws, as well as the interaction of each instruction with all of the other instructions.

6. The "column segment" in this case will include two elements: the upper transceiver assembly and the lower transceiver assembly.

can ever meet the project timelines we have in my company by doing this.

The list of "can't be done" rationales is nearly endless. However, I think responses like these are misguided for several very important reasons:

- This kind of limited thinking is guided by expectations that projects will be managed by trial-and-error, design–test–redesign–retest loops—an idea that I tried to debunk in Chapter 2. There's no time in the project timeline because the project timeline is based on fast-to-prototype, test-your-budget-to-death assumptions.

- The reality is that all of the decisions and assessments that will come out of these multiple studies will, in the end, *be made anyway.* Whether these decisions and assessments arise by default (not really done in any conscious way but assumed or blindly ignored), made by customers, or completed via endlessly complicated and expensive post-release design changes, *these judgments will be made* in one way or another.

- A good deal of DFMEA work seems overly complicated, time-consuming, and difficult because practitioners don't understand the power of robust deductive techniques. Once you see how these techniques actually simplify the work in the following chapters—and also how these methods accomplish other things you need to do to complete a sound design—you'll realize how outmoded that kind of thinking is. Further, like any skill, you need to practice to gain proficiency in order to reduce the time and effort required. But if you never start, you'll never get the benefits.

- The assumption that this is "too much to do" is also grounded in many engineers' long experience with DFMEA, experience that says the methodology is an after-the-fact, check-the-box, fill-in-the-form affair that adds nothing to the design process. If you really believe this about DFMEA and you aren't willing to reconsider that view, I really suggest you stop reading and put this copy of the book up for sale on an online site. The rest of this book won't be convincing or worth your time and effort to read. Your beliefs will be the basis for a self-fulfilling prophecy.

- Finally, I'm all but certain that engineers and managers at NASA felt the same way before the Apollo I disaster—as did Ford before the Pinto program and Airbus before the A380 project ran aground.

Of course, not all projects are as complicated as a mobile phone. Let's look at a simpler, purely mechanical example next.

An Even Shorter and Less Complex Example: Suspension Trailing Arm

Let's consider a simple mechanical design from the automotive world, a trailing linkage arm in an axle. A trailing linkage is often used in the rear suspension of small, inexpensive front-drive vehicles. In this kind of suspension, two linkage arms, one for each wheel, are connected to the rear axle and to the unibody frame. This controls the motion of the rear axle relative to the vehicle frame through pivot joints on both ends of the arm.

It's really not complicated—the arm simply guides and limits up-and-down motion of the axle relative to the vehicle frame, through a limiting arc-shaped path. The coil spring absorbs upward force when the axle moves, while the shock absorbers dampen downward motion when the spring rebounds.

A fairly sophisticated version of this arrangement, with the coil spring and shock absorber combined into a device called a "MacPherson strut," is shown in Figure 3.3.

Now, let's assume that, for a new axle, the forged trailing arm is going to be replaced with a reasonably simple weldment as a cost reduction.

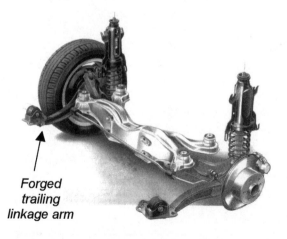

*Forged
trailing
linkage arm*

Figure 3.3   A rear-trailing arm axle, with forged trailing arms and MacPherson struts.

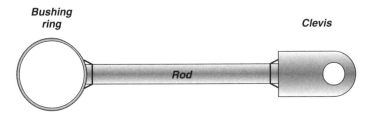

Figure 3.4   Trailing arm concept sketch.

Our task is to design this arm, and the simple concept sketch—the primary input for any new DFMEA study—created for this new design consists of three major pieces: a cylindrical bushing ring, a cylindrical rod, and a U-shaped clevis (see Figure 3.4).

This design is very simple—the ring, which will hold a press-fit rubber bushing, is welded to the rod on one end. The bushing and ring will form the pivot for attachment of the arm to the vehicle frame. A clevis, which will permit the arm to pivot around the axle in the vicinity of the wheel system, is welded to the other end of the rod.

From a system standpoint, this weldment is just one element in the rear suspension of the vehicle. The trailing arm subassembly will include an anti-corrosion coating, probably an e-coat,[7] a molded bushing will be pressed into the ring, and an identification label is also required for the part to be tracked through vehicle assembly operations.

So, to construct a block diagram for this component, we need to show how the weldment is conceptually related to other elements in the rear suspension system. There's certainly a hierarchy involved in this design, and a good block diagram should reveal everything that we need to know at a general level about the weldment so that we can move ahead with the DFMEA study of this new design.

In this case, the linkage arm is not a stand-alone element. Instead, it's only one piece in a larger system. So the starting point for the block diagram won't be the arm itself, but will be "upstream" from the weldment, in the system sense.

---

7. An "e-coat" is a type of water-based paint that is applied by dipping the part in paint or spraying the part with paint while electric current is conducted by the metal part. The resulting electrolysis of water in the paint causes an unusually good bond between the paint and the charged metallic surface.

At the broadest perspective, we are interested in the rear vehicle suspension, and so we'll call this the *system*. (Of course, one engineer's system could be another engineer's subsystem and vice versa—it all depends on your position in the development hierarchy.) Then, the rear suspension can be subdivided into kinematic elements and chassis elements, such as the wheels, tires, and other miscellaneous hardware.

The kinematic elements—those that move to control handling and ride—would include the axle, the MacPherson strut, and the trailing arm, among others.

Of course, there are many ways to divide the rear suspension system, just as there are many ways to divide a clamshell mobile phone.

Once again, the way this system is divided could also have important implications for how the project is managed, including the way the supply chain is managed, how design work is subdivided, and how responsibilities are allocated or assigned in the vehicle development process, including subcontractor or vendor responsibilities.

With all this in mind, a simple block diagram for the weldment could look like Figure 3.5.

Note that there are a good number of items that are missing—we haven't included brake lines (they're in the "other elements" in the chassis subsystem) or many of the axle components and parts. In addition, the dividing line between "subsystem," "component," and "part" is somewhat

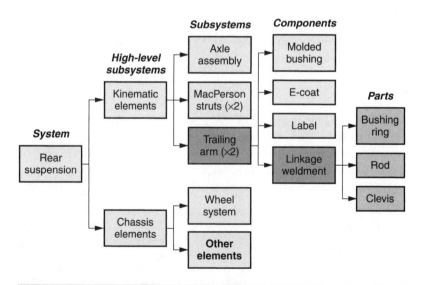

Figure 3.5　Trailing arm weldment block diagram.

arbitrary—but once again it depends on the design philosophy and project management approach that's used in the organization or supply chain.[8]

Once a block diagram is complete, each column segment can be assessed for function. In an ideal world, this would be done using a top-down approach, but in many supply chains, the golden rule applies (he who has the gold makes the rules), and so high-level design information is often not transmitted to those working at the most detailed level.[9]

In this example, we can now see how our project, the linkage weldment, fits into the rest of the rear suspension system. If there were software or electrical elements in this system, they could be included as well—but this is a purely mechanical system, so the elements are mechanical.

The resulting column segment for the weldment has only the three elements that are clearly visible in the concept sketch—the bushing ring, rod, and clevis. Of course, we'd love to have information about the system-level interactions between the kinematic and chassis elements, as well as information about the subsystem interaction between the axle, the MacPherson struts, and the arm assembly, but the reality for most parts suppliers is that the vehicle system engineer is unlikely to provide FMEA results for these interactions.

So, we'll have to consider this as we move forward in our design work—and we will, using a rigorous deductive approach.

Special Team Issues: Alignment with Project Complexity

In Chapter 2, I suggested that team size for DFMEA—and, by implication, team size for engineering project work—should be limited. It doesn't take much working experience, though, to realize that the cell phone project is far too large for a team of six people—no matter how capable they might be. Unless the project is a simple modification of an existing design, dozens or upwards of a hundred or more engineers, quality specialists, and other people are likely to be involved.

---

8. Typically, a component is the smallest unit of a subsystem that can be removed for service—and components are made up of parts, the smallest fabricated element of a component. But some components have only one part while others may have dozens. These definitions are flexible, but establishing some sort of hierarchy or nomenclature for these terms is an important part of managing a development program.

9. This isn't a particularly great way to do things, but it does reflect reality in many, even most, automotive supply chain arrangements.

Conversely, a simple project may well have less work than a single team needs to keep everyone productive.

Fortunately, the block diagram technique I've just described shows a clear way to divide and sort this work into manageable teams, or teams with three to six people. Each column segment is a task for a team, and if a column segment seems too simple for one team, then three possible responses can be considered:

1. One team can be assigned responsibility for more than one column segment—assuming, of course, that the segments selected and team skills result in an appropriate match. Or, one team can assume responsibility for a wider left-to-right portion of the diagram and project.

2. Team members can be assigned to work on more than one project, with different block diagrams, at the same time.

3. The block diagram can be revisited, and the modularity of the design can be changed.

Looking at the trailing arm example, it's highly probable that one team could be assigned complete responsibility for the trailing arm assembly, including the rubber bushing, the weldment subassembly, the e-coat, and the label. Nevertheless, the DFMEA study should be divided into two parts, one for the assembly-level issues and one for the weldment itself, which has a nontrivial column segment of its own.

On the other hand, some portion of a project, like the main printed circuit board on the clamshell phone project, will be too complex for one team. There are again two basic options for aligning the team with the work to be done:

1. The column segment can be subdivided, with one team working on each portion while retaining a common team leader. In this case, a team might have more than six members, but the real work would largely be done in smaller groups. Nevertheless, from time to time a group larger than six might find it necessary to work together.

2. The modularity of the design, as reflected in the block diagram, can be refined so that further subdivision of the complex column segment can be created. This may well improve the design, but the vagaries of individual projects may cause managers to prefer the previous approach, which might be called a *compound team,* instead.

Finally, the overall hierarchy of team members in a complex project needs to be considered. How will the left-side column segments—the larger-perspective, system-view work—be handled?

If the logic of the block diagram is followed, these issues will be resolved early in the project, with individual team leaders from the detail-level column segment teams working together to conduct these initial high-level analyses. These can be done in parallel—during the same general time frame—with early DFMEA work at the detail level. However, these analyses should be relatively straightforward to complete and should not require the deep-dive attention to detail that the right-side column segments probably call for.

In any event, there are some important issues that block diagrams drive that go beyond the basic DFMEA technique that is our primary focus in this book. No block diagram should be considered unchangeable; if this work is done as a paper-based study before significant design resources, prototype hardware, or testing expense begins, then a variety of possibilities can be quickly evaluated—and the possible staffing and budgeting consequences can be quickly worked out.

If this kind of preparation isn't carried out, the project may well drift in the early stages. As these issues are sorted out, there will be uncertainty and concern sown in the minds of the engineers working on the project. Nothing tends to cause working-level engineers, designers, and quality staff to lose confidence in management faster than ever-shifting ideas about how work is divided and responsibilities are assigned, and the specific technical goals for a project.

While the division of responsibility (and assignment of authority) will never be clear enough to satisfy everyone, block diagrams are great tools that allow managers to think through the consequences of alternative project management schemes.

If you are willing to work through this approach once or twice, you will quickly see how powerful this tool really is.

DEVELOPING "FIT FOR USE" STATEMENTS OF FUNCTION: ROBUSTNESS AND P-DIAGRAMS

Too many engineers—and too many companies—take a simple-minded approach to design: "Give me the specs and I'll design a product to meet the specs."

However, that's not a sound approach in the quality-obsessed markets of the twenty-first century. No set of specifications is ever fully complete, although a set that's derived from a comprehensive function analysis (see Chapter 4) can be closer to complete than an arbitrary set of requirements derived from history or other sources.[10]

The real standard—the standard for world-class design—is to develop products that are *fit for use*. What does this mean? It means that the product will not only work properly in conjunction with other products (which can be seen in the block diagram), but it will work properly in the overall environment it's designed for and that it will operate properly when used by real human beings.

In other words, a product that is fit for use will operate properly in a reasonable set of circumstances when used by the buyers that are expected to purchase or use the product.

A product that is fit for use—within the boundaries that match your assumptions about the customers you expect will buy the product—will also be *robust*. In other words, it will be strong enough, durable enough, and damage-tolerant enough to be used in most (hopefully all) expected environmental conditions and usage situations. It will also work well with other bits and pieces if the product is part of a larger system.

Would a better understanding of robustness have headed off the Apollo I fire? Would the interaction of the electrical system and atmosphere have been identified? After you've read the rest of this chapter, I think you'll conclude that it would have been difficult to miss this.

Now, it does take some effort to get a handle on these factors. For example, let's consider a power brick—the AC-to-DC converter that is used with most laptop computers. Let's start out with some questions that will help us understand the environmental and user factors.

Should the power brick be expected to operate in a saltwater environment? Probably not—unless you are designing a portable computer that is suitable for use on an oceangoing vessel. Should the brick work when exposed to dust? How much? How much humidity is acceptable? What ambient temperatures are acceptable? In Kuwait, temperatures can sometimes reach 125°F (50°C). Will your product need to operate in this environment? All of these questions are important to defining the limits of robust performance for a design.

---

10. Quality function deployment (QFD) can get close—but that's just another way to derive functions. And QFD usually isn't as comprehensive as the methods outlined in Chapter 3 and Chapter 4.

What about usage factors? Should the brick be resistant to a human stepping on the casing? How big a person should be considered? Can the brick be dropped without failing? How great a drop should be acceptable?

And we can't forget that the brick must operate with other elements in a system. This will include an AC power outlet, the laptop system the brick is intended for, and the cords that duct power from the AC outlet to the brick itself.

Whew! This is a relatively simple product and yet there are probably a dozen or more factors that need to be considered. It could very well be that the historical standards that you've developed will cover these issues, but it's equally likely that you'll miss something that will cause problems.[11]

On the other hand, there's a practical limit to what you should consider. Dropping the power brick from a three-story window doesn't seem reasonable, nor does subjecting the brick to $-100°F$ $(-75C)$ temperatures.

Sometimes, design teams fall into one of these traps. At one time, one of the major automakers actually required that trunk, or boot, carpeting must be capable of prolonged exposure to sunlight without fading. That requirement added a meaningful cost to the carpet. How did this happen?

As the story goes, a company executive had possession of a vehicle for an extended evaluation period and drove around for several days in the middle of summer with the trunk open, as he hauled around some antiques he was buying and selling. The carpet faded and, as can happen in an organization, the executive made sure that this wouldn't happen again.

So, what was the net result? First, all trunk carpeting soon had to meet a new UV exposure standard. This raised the cost of the carpeting—and the price of the vehicle had to be increased to cover this additional bit of robustness.

Now, there's little doubt that some consumer, somewhere, will drive around in bright sunlight with the trunk open. How many might do that? A few? Thousands? At some point, this makes no sense, because the automaker would be asking most of their customers to subsidize the behavior of a very few people, customers whose behavior is so outside the bounds of reasonableness that the basic environmental factor—UV exposure for trunk

---

11. That's particularly true if you are introducing a new design or significant new technology into an existing design. This is something that's dogged Toyota for years: their designs are wonderfully robust when they are based on long historical baselines. However, they tend to have many annoying and even awful aspects when Toyota has no baseline to compare against.

carpeting—could reasonably be deleted from the list of "fit for use" factors that are included in the design criteria.

If you understand these issues, then you can develop a list of measureable functions that will drive a design to a fit for use conclusion.

A simple tool that can help you derive the driving factors is called a *parameter diagram, or P-diagram.* P-diagrams are also important in setting up design of experiment studies and offer additional insights into important design considerations that go beyond function. However, a P-diagram also offers a tool for organizing your assessments of fit for use factors.

So, let's take a look at how you would go about constructing a P-diagram, in its fullest form, while recognizing that only a portion of the results will be useful for our efforts in function analysis and DFMEA.

The goal of a P-diagram is to relate the general performance of a product to "noise factors" that can interfere with performance.

For almost any product, there are factors that are applied to the device, or, in the case of software, data inputs. We call these "signal" factors. The device or product then needs to respond to these signal factors. Responses that are within the required limits then frame a statement of "ideal function."

For the power brick, AC power is a signal factor. This power, both 110 and 220 V inputs, with appropriate limits to voltage and current, is then transformed into DC power, again with some appropriate limits. As long as the brick performs within these limits, we say the product is performing the ideal function.

Graphically, this looks something like Figure 3.6.

As long as the signal factors stay within the anticipated limits, we expect that the product will perform in a way that will remain within the limits of the ideal response. For a power brick, this means that as long as the AC current entering the brick is within expected limits, the DC power leaving the brick will be within limits that are compatible with the laptop for which the brick was designed.

Or will it? There are a number of factors that are "pushing" on this ideal functionality, and we call these "noise factors." So, our diagram just became a bit more complicated (see Figure 3.7).

Figure 3.6   The ideal function statement—the foundation of a P-diagram.

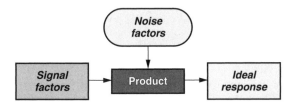

Figure 3.7   The ideal function can be affected by noise factors.

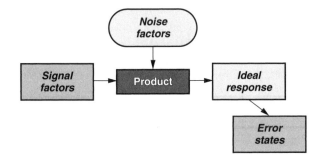

Figure 3.8   Noise factors can cause the ideal function to be disrupted.

If any noise factor presses hard enough, the product's performance may fall outside of the ideal response, creating what is known as an "error state" (see Figure 3.8).[12] And there may obviously be more than one error state.

The next step will be to add specific noise factors. As Genichi Taguchi has suggested, noise factors can be grouped into five different categories:

- *Environmental interactions.*

- *Changes with time or usage.*

- *Customer use factors or life cycle issues.*

- *System interactions.* These factors should be visible in the block diagram.

- *Variation within specification limits.* This is the impact of multipart tolerance stack-ups, which will include not only mechanical size tolerances but also material property variations.

---

12. Error states are effectively the same as failure modes, although the statement of a failure mode is more demanding than the description of an error state.

If you think there's potential overlap between these categories, you're correct. Getting an issue in the correct "box" is not as important, though, as identifying all of the pertinent noise factors. If you aren't sure which category is the most appropriate, make a decision and see how that works. If it doesn't make sense, you can always change this later. The specific "binning" of factors into correct categories isn't as important as getting a comprehensive list of factors that the product must interact with in some way.

The goal is to get a broad *and* deep insight into how the product works in the real world. Because real-world issues can often get a bit messy, a P-diagram may also be a bit untidy. If you practice this technique, though, you will find that succeeding diagrams take on greater clarity and utility.

When we add in these noise factors, the P-diagram is close to being finished (see Figure 3.9).

To complete the P-diagram, we need to add one more item. Hopefully, our design will be evaluated to ensure that the noise factors, as long as they remain within anticipated limits, will not disrupt the signal-to-response chain. That's another way of saying that our design will be robust, because we don't expect the product's performance to vary sufficiently from the ideal response to enter into an error state, even in difficult but defined conditions.

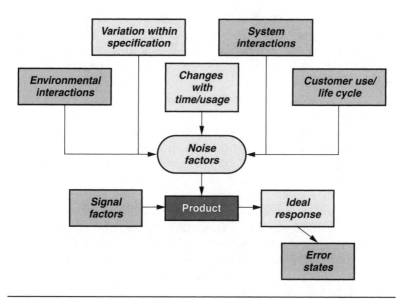

Figure 3.9 The P-diagram structure with all noise factor categories.

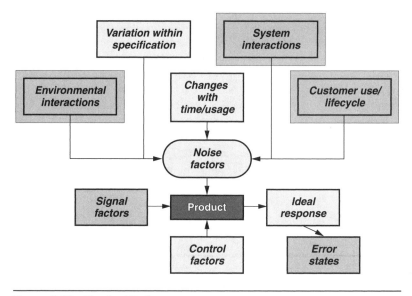

Figure 3.10 The final P-diagram structure.

We call these evaluation activities "control factors," and they become the basis for design verification.

The three *highlighted noise factors* (see Figure 3.10) are central to functional analysis for DFMEA: environmental interactions, system interactions, and customer use/life cycle considerations. For each of these factors, we need to compile an inclusive list of the expected "noises" that will fall under each category.

The other noise factors serve important purposes in extending robust design methods, but we don't need to discuss them to conduct DFMEA studies. Nor do we need to know at this point what the control factors might be or what the error states might include.

Returning to our trailing arm weldment, we can work through each of these noise factor groups, starting with system interactions.

But first we need to decide how signal factors and ideal responses can be described. The trailing arm, as a mechanical component, is driven by energy from vehicle motions. The energy is applied to the arm in the form of force and motion. And what does this component do with the energy? It absorbs very little of it and largely transmits this back to the chassis system through the bushing. Of course, there are limits for each of these (both on the signal side and on the response side of the diagram), but that's not decisive at this point.

If we look at the block diagram in Figure 3.5, we can see several system interactions:

- There's interaction with the wheel system, where clearance is important.

- There's interaction with the molded bushing, e-coat, and label, which make up the remaining elements of the trailing arm.

- There's interaction with the chassis system, which locates the bushing ring.

- There's also interaction with the axle, which transfers force and motion to the linkage—and also dictates the size and location of the linkage arm subassembly through the clevis.

What are the customer usage and life cycle factors? The way the customer drives will have a very important effect on design, and this will occur primarily due to the jounce position (the farthest limit that the axle can travel upward) and the rebound position (the travel limit in the downward position).

Finally, the environmental factors must be listed. For the weldment, the two critical factors that must be considered are stone impingement, which can damage the weldment, and salt water, which can cause corrosion.

Putting this all together gives us an abbreviated P-diagram for the weldment (see Figure 3.11).

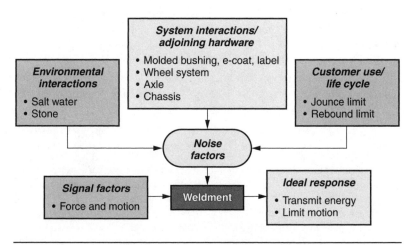

Figure 3.11 Abbreviated P-diagram for trailing arm weldment.

As a final note, there are some things you need to keep in mind when building the P-diagram. The diagram will only be as good as your actual knowledge of the noise factors. If you have access to historical information about failures, warranty claims, and customer complaints, you will probably have good information about usage and life cycle factors. Similarly, a keen understanding of the way a component operates will make the system interactions clear—although a good block diagram also aids comprehension.

At some point, though, particularly with new designs and technology, you will need to depart from deductive thinking and dip into your personal or organizational experience to identify all of the likely and reasonable noise factors. How will a product be used? What environmental conditions must your product endure? (System interactions are likely to be visible in the block diagram set, limiting the need for imagining these noise factors.)

Whether you have learned this from historical data or from intuitive understanding of usage patterns, it's critical that you know what the external environmental factors might be. For example, automotive gearshift mechanisms, particularly floor-mounted shifters, must be resistant to coffee and carbonated beverages—because these liquids are regularly splashed or even spilled into the mechanisms.[13]

If this leads you to suspect that creating the P-diagram is the weakest link in function analysis, you are correct. No one has yet devised a deductive tool to accomplish this, although it's certainly possible to do this using some form of extended database and search engine. But, even with an automated database, the technique will be weaker as a rational and deductive tool than any other step in the DFMEA process.

The best antidote for this limitation is to gather as much historical information as possible—and then make sure that the P-diagram is created by a properly constituted team who knows and understands the usage of the product. However, you do need to watch for engineers who think that everyone is just like they are and that consumers or buyers will use the product the same way they might. You need to put yourself into your intended customers' shoes, particularly if your intended customer is very much unlike the people on the DFMEA team.[14]

---

13. All of these issues will emerge as we progress through the DFMEA process, however. This is also a good example of how the noise categories sometimes blur. Is cola in the shifter a usage factor or an environmental factor? Again, it doesn't matter as much as capturing the basic issue.

14. For example, it can be difficult for young engineers to understand what a product intended for senior citizens might have to be like to be robust. In the same vein, many automotive engineers think everyone is—or perhaps should be—a gearhead.

Nevertheless, if block diagrams and P-diagrams had been created for the Apollo I capsule, I find it difficult to believe that the interaction of the wiring system and the atmosphere would have been dismissed. Similarly, the interaction of the axle, fuel tank, and crash dynamics would have been harder to overlook in the Ford Pinto program.

STARTING THE MAIN WORKSHEET

If you've completed a block diagram set and then prepared a P-diagram for the column segment of interest, you can quickly complete the top section of any DFMEA worksheet.

There are many different DFMEA forms in existence, and the top portion, often called the "header" section, is the first part of the form that requires attention. However, this is really just a summary of scope information and really includes little or no analysis to complete.

On the other hand, it's far better to have completed a block diagram set and relevant P-diagrams before completing a DFMEA header section. If you understand the true scope of the project, completing most header sections is simple and straightforward.

And, because there are so many different kinds of headers in use today, there's really little benefit in showing an example header. Header sections usually include two kinds of information: product scope information and internal project management facts.

Product scope data usually includes identification of the type of DFMEA being considered—system-, subsystem-, or component-level study. It will always include a description of the specific item being studied and perhaps some identification with higher-level systems. For example, in the automotive world, this would require identification of the model year and vehicle in which this system or component will be used. Scope information will probably include a part number or other alphanumeric identifier that will be used to track and coordinate design information.

Project management information can be quite varied and depends more on the way the business or organization conducting the study manages projects. Common project factors include DFMEA team member information (names, contact information), the project leader or manager's name, the name of the design-responsible engineer (who may or may not be the project leader), date information about the study, including the date of the study, the target date for release of the design, or other information.

Header sections also typically include information such as the name of the person who actually prepared the form, a cross-reference number for the DFMEA document (which may be different from the part number) and relevant supply chain information.

Finally, a good header section will also include basic change control information, such as the initial completion date for the study and a record of dates when changes were made to the study after the initial completion date. However, a detailed change record usually is not included in the header itself; it's often added as a separate sheet or section in the DFMEA record, and it resembles the change record often shown on engineering drawings.

None of this is really complicated, so you should be able to complete a header with just a minimal knowledge of your own organization's requirements. But, knowing details about the scope of the project, particularly information developed and tracked with block diagrams and P-diagrams, will be very useful in generating concise and accurate header entries.

4

Step 2—
Understanding
Function

After the scope of a project is clearly understood, a properly staffed team is ready to undertake the most pivotal aspect of any design project: developing a deep and wide-ranging understanding of the functional requirements for the product being developed.

To the greatest degree possible, this should be a detailed description that comes as close to being all-inclusive or complete as possible. Of course, no functional description is ever perfect, but you'd really like to be as close to perfect as reasonable amounts of time, money, and intellectual energy permit.

To accomplish this, we need to develop two additional intermediate outputs: an interface matrix and a comprehensive listing of initial or proposed specifications (see Figure 4.1). However, before we can really tackle these items, we need to take a look at what it really means to define and delineate function in any product.

If your goal is to get meaningful results from DFMEA, it is critical—absolutely vital—to define and describe product function before attempting to determine what error states or failures might arise. The fact that so many people start any FMEA by brainstorming failure modes is a demonstration of how severely the underlying logic of FMEA has been corrupted by

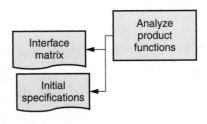

Figure 4.1 Step 2 of DFMEA process.

ineffective techniques. Even more, it is a serious indictment of how poorly FMEA has been taught over the past six decades.

With Step 2 in the DFMEA process, we will also start using the core portion of the DFMEA form—the columns shown on the form. Perhaps the most common mistake that people make is completing a DFMEA by working on a row-by-row basis.

If you work through the DFMEA process row by row, you'll fall into several traps that are almost certain to make DFMEA a tedious, time-wasting exercise with minimal value. The things that most often happen include:

- Great detail and thought goes into the first few rows. As more rows are added, the detail and thought diminish and, by the end of the process, the tail end of the study is often empty of insight and predictive power.

- The use of the rating tables, which we will consider in Chapters 6, 7, and 8, will be inconsistent and problematic. Teams will likely suffer through repeated or nearly endless arguments about the use of the tables. Using the tables will also be time-consuming—and the time used won't add much (if anything) to your understanding of the design.

- Cause–mode–effect chains will be confused. Effects and causes will be mixed up, modes won't be modes, and—if effects and causes are reversed—the entire study will be nearly worthless.

- Similar cause–mode–effect chains will be described in very different ways, leading to confusion about the underlying design intent, and methods of verification will probably be weak or even misleading.

So, you need to learn to say it over and over again: "Work column by column . . . work column by column . . . work column by column . . . ." And, once you have that burned into your brain, remember that you should *complete* each column on the form before moving to the next column on the form.

Ideally, you will limit this rule so that you complete each column segment from the block diagram set. This really means a "column by column within a column segment" principle:

- Choose one column segment from the block diagram set.

- Work through the DFMEA process and complete the form column by column for the first column segment selected.

- Move to the next column segment in the block diagram set.

- Work through the DFMEA process column by column for that column segment.

THE LANGUAGE OF FUNCTION

What is function? If you look up this word in *Webster's New Collegiate Dictionary,* you'll find this definition: *"the natural, proper, or characteristic action of any thing."* This is very close to a working definition that we can use for DFMEA, but there are certainly other aspects of function that may play a role in understanding what a product must do to satisfy marketplace needs.

Hyperdictionary.com, an excellent online source, offers ten possible definitions, including "what something is used for; 'the function of an auger is to bore holes'" and "a relation such that one thing is dependent on another; 'height is a function of age.'"

These are all very good and are actually quite close to the mind-set we need to have for understanding function in a product. However, we need a very specific operational definition for function to move forward in DFMEA. For our purposes, we will say that:

> Function is a *description of the purpose of a product, constrained within the limits of expected usage and system interactions.*

In order to develop a list of functions that is sufficiently detailed and realistically comprehensive, we must first stop and recognize that FMEA is a linguistically based activity. The underlying meaning of results that are obtained from any FMEA study will be significant only if the language and grammatical construction of function statements are clear and unambiguous.

After all, if you can't clearly and unambiguously describe what a product is supposed to do, how can you execute a design that will do just that? How could you understand what could go wrong if you don't have a coherent understanding of what is supposed to go right?

Moreover, as we will see in Chapter 5, constructing reasonable and powerful mode statements depends heavily on using a disciplined and rigorous method of describing function. Even more to the point, to maintain consistency and accuracy in describing cause–mode–effect chains, it is essential that a coherent and enduring grammatical construction be used when describing function.

To properly describe function in a way that is grammatically consistent as well as powerful, we must use an active verb and a measureable noun for each function. Adjectives may be required, particularly to differentiate how different elements of a design are related to other elements, but at the heart of any good function description is an active verb and measureable noun combination.

What do we mean by an "active verb?" An active verb is a verb that avoids the passive voice in a statement of function, and over the years, I've found it's easier to explain what you must not do rather than explaining what you must do to use an active verb.

In the passive voice, a form of the verb "to be" is followed by a past participle, a verb that usually (but not always) ends in –ed and puts the action in an earlier period. For example, "force was transferred by the beam" is passive. To restate this in a more active way, you can say "the beam transfers force."

Passive construction in a clause or a sentence tends to be vague and indistinct, and if you have a vague and indistinct statement of function, you'll have a hazy and imprecise statement of mode. You will also end up with a function that is difficult to verify and will allow a wide range of interpretation when you analyze or test your design for that particular functional performance criterion.

In short, passive construction in a function statement often leads to a lack of rigor or even a vague level of understanding about the product's design requirements. And a lack of precision in a function statement will lead to imprecise cause–mode–effect chains.

Unfortunately, engineers *love* passive statements of function. It leads to "wiggle room" in interpretation and leaves plenty of openings for changing what a design is supposed to do. It's a major enabling factor in the test–fail–redesign–retest–fail again loop that we know leads to budget busting projects that are late to market.[1]

To avoid using passive function statements, you need to practice, practice, and practice some more. However, when you are starting, you must—to the greatest degree possible—avoid the so-called "nerd verbs." These are verbs that almost always lead to passive function statements—statements that reduce the power of DFMEA.

The most common "nerd verbs" include:

- Provide

---

1. In fairness, young engineers are usually exposed to scholarly papers that make extensive use of the passive voice; this lends an air of authority to passive sentence construction and makes a passive construction seem more respectable and significant—not unlike the construction of this footnote. . . .

- To be (and all of the tenses of this most common verb)

- Supply

- Facilitate

- Allow

When I'm working with a group of engineers who've never been exposed to sound FMEA processes, I find that these nerdy verbs are, more often than not, the verbs of choice. Then, when I engage these engineers in dialogue, they frequently admit that they prefer these descriptions because these verbs permit a "flexible" description of product performance.

This means that specifications can be less confining, test results can be subjected to interpretation that leads to acceptance, and the hard details of satisfying ever more demanding customers can be left to the manufacturing people.

If you are, like so many tech heads, a fan of nerd verbs, here is a simple way to overcome this habit. Turn the noun into a verb. For example, "allow clearance between X and Y" is a common way that an engineer might express a mechanical system condition. This allows a great deal of room to maneuver when discussing a design. Does clearance mean "not touching" or does it mean "not interfering"?

To fix this, we could restate this relationship as "clear X relative to Y," which would be active and more precise. By doing this, we've changed the verb "clearance" into an active verb "to clear," and the result is superior to "allow clearance."

With practice, though, you can develop skill with other active verbs that are more direct and specific. For example, you could say "limit clearance of X relative to Y," a verb–noun statement that draws attention to the need to define the specific requirements that a designer must achieve. If clearance is kinematic rather than static, you might say "maintain clearance of X relative to Y over the entire range of motion."

In the end, a sound function statement is one that provides a clear and measureable purpose for the product you are designing. This is true whether the product is mechanical, electrical, or even a software system. If you can't describe the function in a measureable way, you may not be engineering a product—you might be creating art.[2]

To make sure you understand this, let's consider a very simple example: a mechanical pencil, the inexpensive type you might receive as a promotional

---

2. My colleague Kim Pries has observed that great artists also achieve great clarity; but, in my experience, most engineers and technical people are rarely great artists.

giveaway at a trade show. What is the purpose of this object? Most people would respond, "to write." First, that's an infinitive, not an active construction, and, even more importantly, that's not a product function.

If you can invent a pencil that writes, you might well become rich beyond your wildest dreams.

Instead, the purpose of the pencil could be described as "transfer graphite to paper" or something similar. Is that measureable? Yes—the width of the resulting marks could be measured, as could the length of total transfer in the life of one lead insert, and even the darkness of the marks.

Of course, there are other functions in a mechanical pencil besides "transfer graphite" that would probably be seen in a statement of ideal function in a P-diagram. To get a more complete list of functions, we need to use another deductive tool.

PUTTING THE PIECES TOGETHER: THE DEDUCTIVE INTERFACE MATRIX

Now we are ready to use the information developed in the block diagram and P-diagram to set up an interface matrix. By properly applying interface matrix techniques, we will be able to develop statements of function for a product by a deductive method rather than by brainstorming, imagination, historical baseline, or some other inductive method.

Many different types of interface matrices can be created. The particular type that is most useful for deducing the function of a product includes these features:

- The matrix relates the internal elements of a project to each other.

- The matrix relates the internal elements of the project to the three critical noise factors identified in the P-diagram.

- The interaction between each interrelated element or specific noise factor is explained through statements of function—active verb and measureable noun statements.

We can start the interface matrix by returning to the block diagrams that were described in Chapter 3. For the suspension trailing arm, the project scope—our direct design responsibility in this example—is for the weldment. Extracting this portion of the diagram, we get a simple three-element column segment, as shown in Figure 4.2.

This column segment contains all of the components or parts that we are interested in and nothing more. Of course, this is a very simple example, and there are only three internal elements in the weldment—the ring, rod, and clevis.

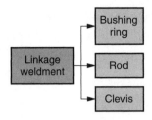

Figure 4.2   Portion of block diagram for suspension trailing arm.

It would be possible to look at function at a higher level, and this would be accomplished by simply going upstream in the block diagram set, picking a subsystem, and then extracting the column segment that derives from the subsystem element.

For example, if your design team has responsibility for the complete trailing arm and not just the weldment, you would have four elements: the molded bushing, the label, the e-coat, and the weldment itself.

In this case, you would treat the weldment as one element and you wouldn't try to understand the function of the individual parts (ring, rod, and clevis). That doesn't mean that the function of the individual parts isn't important, but it does mean that the weldment would have a separate column segment in the block diagram as well as a separate interface matrix—and a separate section in the DFMEA itself, following the "column by column within a column segment" principle.

This is also a good time to take another look at the block diagram itself. If you find that a column segment has as many as a dozen elements, you may want to rethink how you have subdivided or modularized the design. It may even be necessary to change the block diagram and associated P-diagram.

Why? Well, if you have a column segment with 17 elements, you will need to understand, at a detailed level, how all 17 elements interact with each other. That's a total of $(17^2 - 17) / 2$ direct, internal interactions[3] or 136 total interactions. That's complicated, and troubleshooting such a system is very difficult, if not impossible.

That doesn't mean that dividing 17 elements into subsystems or components will mean less complexity. However, by subdividing a design into manageable subsystems, components, and parts it is possible to deal with this complexity. This occurs because "grouped" elements can be visualized or assessed as a single element, and it won't be necessary to look at all of the possible permutations of interaction.

---

3. The actual count for any matrix is $(N^2 - N) / 2$.

For example, we do need to understand how the weldment interacts with the wheel system. But do we need to know how the weldment interacts with the lug nuts, rim, tire, valve stem, and valve cover—on an element-by-element basis? No—we just need to know how the weldment interacts with the wheel and tire, or wheel system.[4]

It is important to realize that using this approach doesn't make interactions go away. Instead, it allows you to treat many of the system and subsystem interactions in a way that is easier to understand and, to a first approximation, to bundle these interactions so that you are not overwhelmed by minute (and generally insignificant) details.

The more that you work with block diagrams, the more you will understand that the way a system is modularized is a critical design decision. Very long column segments should be seen as a red flag suggesting that a different division scheme should be considered.

For our current example, though, we have deliberately kept the complexity low; with just three elements, we have just three interactions to consider.[5]

Since this is a "matrix," it must have rows and columns. For the "internal" portion of the diagram, you will create identical rows and columns, with one row and one column for each element. In our case, this will be a row and column for the ring, the rod, and the clevis. So, you should start by constructing a simple matrix with headings for the rows and columns as shown in Figure 4.3.

We don't need to concern ourselves with the interaction of the rod to the rod, or ring to the ring—and so these cells in the matrix are called *identity cells*.[6] And, because there's no "directionality" in the interactions (the interaction of the ring to the clevis is no different than the interaction of the clevis to the ring), there are cells that are effectively duplicates. We call these cells *mirror cells*.

---

4. This can become extremely complicated in software development, particularly where object-oriented programming (OOP) is used; with OOP techniques, you can have hundreds of discrete elements, each of which can interact with another. This may be one reason why OOP can sometimes be thought of as "oops."

5. If a part has only one element, say a one-piece brake rotor, then there will be *no* direct internal interactions, since $(1^2 - 1) / 2 = 0$. However, there may be dozens of external interactions.

6. There are interesting issues that arise in identity cells, particularly for a matrix that is derived from an upstream column segment in a block diagram. Conceptually, identity cells actually contain subordinate matrices for column segments that are downstream in the block diagram. But that's a bit of geekiness that we don't need to worry about in most cases.

Internal element	Rod	Bushing ring	Clevis
Rod			
Bushing ring			
Clevis			

Mirror cells

Identity cells

Figure 4.3 Internal interface matrix for suspension trailing arm example.

Internal element	Rod	Bushing ring	Clevis
Rod			
Bushing ring			
Clevis			
External element	Rod	Bushing ring	Clevis
External hardware			
User issues			
Environmental factors			

Mirror cells

Identity cells

P-diagram noise factors

Figure 4.4 Internal interface matrix for suspension trailing arm example with general groups of noise factors added.

However, this internal portion of the interface matrix isn't the whole story. Not only are we interested in the interaction of each element with the other elements, we are also interested in the interaction of each element with each important P-diagram noise factor—the environmental factors, the user issues, and the external hardware factors.

So, we need to add more rows. We don't need to add more columns, because we aren't concerned about the interaction of the noise factors with each other. This can often cause many, many rows to be added, depending on the design. But, in general, there are three groups of rows to be added: external hardware, user issues, and environmental factors (see Figure 4.4).

So, for our weldment assembly, let's now replace the three general noise factor categories with all of the specific noise factors from our P-diagram for the weldment to the matrix (see Figure 4.5).

Now we've created an interface matrix. The next step is to populate the cells—and this is done by asking a very simple question:

What is the interaction between the row heading and the column heading?

Internal	Rod	Bushing ring	Clevis
Rod			
Bushing ring			
Clevis			
External	Rod	Bushing ring	Clevis
Wheel system			
Axle			
E-coat			
Label			
Molded bushing			
Jounce limit			
Rebound limit			
Stone impingement			
Salt water			

Figure 4.5   Interface matrix for suspension trailing arm example with specific noise factors.

Answer this question with one or more active verb–measureable noun function statements. In all cases, there are four different types of interaction that can occur:

- A physical interaction, which can involve contact and is largely concerned with clearances and fits

- An energy transfer, which can involve voltage and current in an electrical system, or force and motion in a mechanical system

- An information exchange, which typically occurs in software but can also occur in electrical or even mechanical control systems

- A material exchange, which can occur when material from one element is actually transferred to the other element

Again, you don't need to worry about directionality in answering this question. If there's directionality, this should be evident in the statement of function, using the active verb–measureable noun construction.

In fact, if the interaction is two-way, with element A doing something to element B, but element B doing something to element A that is truly different, you simply need to include both statements in the matrix cell at the intersection of element A and element B. Don't try and create an elaborate compound sentence to describe this. Just break the idea down into two parts and create two separate active verb–measureable noun statements.

The main difficulty that most people have with describing these functional relationships is failing to understand that each matrix cell should only contain direct, serial relationships. This can get complicated, so let's consider an example from human anatomy.

Over the years, I've led an active life. As a result of wear and tear and possibly genetic predisposition, I've worn out my hip joints. The hip joint is a relatively simple ball-and-socket arrangement. The top portion, the socket in the pelvis, is called the acetabulum. The acetabulum is lined with cartilage, and the lower portion, the thigh bone, or femur, has a ball on the top of the bone, which fits into the acetabulum.

The cartilage acts as a cushion, and there is a lubricating fluid that keeps the socket working smoothly.

Over time, particularly if you've done things that put a great deal of impact load on your legs, you can damage or wear out the cartilage. Then, you end up with bone-to-bone contact and some nontrivial "sloppiness" in the alignment of your femur to your pelvis. It's very painful and, left untreated, can cause you to progress from a cane to crutches to a wheelchair in a few years.

This started to affect me in my late forties and, because the surgical implants that "fix" this condition have a finite life expectancy of about thirty years, the doctors advised me to wait as long as possible before undergoing surgery to correct the problem.

In the meantime, I was risking damage to my knees. Because the hip joint was chronically misaligned with my leg, the impact load path was often out of line, too. That resulted in odd and potentially damaging side loads on my knees. As a result, I wore knee braces most of the time—and did so until I had both hips surgically replaced.

Now, we all know that "the shin bone's connected to the knee bone, knee bone's connected to the thigh bone, thigh bone's connected to the hip bone" and so forth. If we applied this to an interface matrix, what would it look like—sans noise factors? (See Figure 4.6.)

What's the relationship between the shin and knee? One function statement would be "transfer impact load from shin to knee." Another could be "limit rotation of knee relative to shin." (Unless you are an orthopedic surgeon, you probably can't list them all. But they all matter.)

The same "transfer impact load" will occur in the interactions between the knee and thigh and the thigh and the hip. But, *will there be any direct interaction between the knee and hip?*

Internal element	Shin	Knee	Thigh	Hip
Shin				
Knee	*Transfer impact load from shin to knee; limit rotation of knee relative to shin*			
Thigh	*None*	*Transfer impact load from knee to thigh; limit motion of thigh relative to knee*		
Hip	*None*	*None*	*Transfer impact load from thigh to hip; align thigh to hip*	

Figure 4.6   Internal element functions for the hip joint example.

The answer—at least insofar as we are concerned in function analysis—is no. The direct, serial relationship follows from shin to knee to thigh to hip. A misalignment of the hip transfers load and motion (energy, in a more general sense) from the knee to the thigh and *then* to the hip. So, misalignment of the hip causes the thigh to accept loads at angles that aren't "normal," which in turn applies force vectors to the knees that aren't "normal."

The same kind of thing can be seen in a simple electrical circuit. Suppose a resistor is connected to a capacitor, which is then connected in series with a rheostat. There's no direct relationship between the resistor and the rheostat, even though the rheostat will be affected by the resistor's performance.

In the matrix, you would see the interaction between the resistor and rheostat by understanding the relationships between the resistor and capacitor and the capacitor and rheostat.

On the other hand, if the resistor and capacitor were in parallel and feeding into the rheostat, then you would see a direct relationship between the resistor and rheostat, a direct relationship between the capacitor and rheostat, and a direct relationship between the resistor and capacitor.

If you don't follow this rule, you will find that the interface matrix is terribly confusing and ultimately useless in assessing function. You'll get so tangled in trying to see what-does-what-to-which that you will never get a sound and repeatable result.

Keeping all that in mind, we are now ready to apply these ideas to the trailing arm weldment.

It's usually best to start with the internal elements and work column by column. Then move to the external elements and work row by row.

The interaction of the rod and bushing ring appears to have three functions:

- Transfer force between the bushing and rod (it can go either way and the direction really doesn't matter).

- Locate the rod to the ring (this could be separated into various feature controls, such as angularity, length, or other factors; however, those details are probably better handled in the next major step of function analysis—specification development).

- Retain the ring to the rod.[7]

There are three similar functions between the clevis and the rod—but, in this case, there is a direct, serial relationship between the ring and clevis.

The axis of the clevis must be controlled in terms of angularity relative to the bushing ring. Because the rod is symmetrical in the controlling axis, this can't be stated between the ring and rod and—separately—the clevis and rod. As a result, there is a direct relationship between the clevis and ring.

Once you've completed the internal relationships, you can start on the external relationships. What's the relationship between the wheel system and the rod? There really isn't any, so you should enter "none" in the cell. Entering "none" is not just tidiness; it's important to do this so that later on you know that you made a conscious decision rather than a default decision. If you leave the cell blank, you won't really know whether you believed there was no interaction or you just didn't think about it.

Similarly, there's no real relationship between the bushing ring and the wheel, but there is a relationship between the wheel and the clevis. The clevis must not interfere with the wheel in the various kinematic positions that will occur in operation, so a function of "maintain clearance between clevis and wheel system" is an appropriate function.

That concludes one line on the external portion of the matrix. Keep moving through the lines until you've completed all of the lines associated with noise factors.

---

7. Notice I didn't use the verb "attach." Attachment is something done in manufacturing; the design must simply retain the ring. This may seem trivial, but precise thinking about design-oriented issues is critical to good DFMEA practices.

For the e-coat, the interaction with all three elements in our scope is the same: resist corrosion. Because corrosion occurs frequently, in function analysis it's probably important to point out that "preventing corrosion" is almost impossible. There's a bit of it in almost all circumstances, so don't use the verb *prevent* unless you really mean it.

However, there is another issue that sometimes arises when you complete the rest of the lines in the external portion of the matrix. In this case, what's the interaction of salt water with the ring, rod, and clevis?

If you believe that the presence of e-coat is included in the trailing arm assembly to resist structurally significant corrosion, the direct, serial relationship between the weldment elements and salt water may well be "none." If, however, the purpose of e-coat is just to resist corrosion so that the undercarriage is shiny and clean in the showroom, "resist corrosion" may still be a function that is derived from the interaction between salt water and each of the weldment elements. However, if the issue were one of appearance, then the specification for this material would be different than it would for structural corrosion.

In either case, the impact of salt water would then be found at the next upstream column segment in the block diagram—where e-coat is one element of the trailing arm assembly. You'd then be justified in deleting the row for salt water and adjusting the P-diagram appropriately.

When you've entered "none" in every cell in a noise factor row, then you're really justified in removing this row from the matrix. After all, you've determined that the noise factor really doesn't apply, at least insofar as you've limited the scope of your project. So, with all those ideas in mind, Figure 4.7 shows what the matrix for the weldment has evolved to.

If you look at this result carefully, you will see that the functions shown in the external rows are, in fact, the functions that must be completed successfully if the design is to be robust. By stating this in terms of function, you have an all-inclusive list of all the things the design must do to avoid losing performance when critical noise factors (from the P-diagram) are encountered. This alone makes the interface matrix a worthwhile exercise, but there's an even greater set of benefits, as we'll see.

Developing Skill with Interface Matrix Techniques

A variation on the basic method can be employed, particularly when you are learning this method or when you are constructing a very long and complex matrix. Start by constructing the matrix the same way—rows and columns for each element in the column segment of interest, and additional rows for each relevant noise factor.

Internal	Rod	Bushing ring	Clevis
Rod			
Bushing ring	*Transfer force, locate rod to ring, retain ring to rod*		
Clevis	*Transfer force, locate rod to clevis, retain clevis to rod*	*Control angular relationship*	
External	Rod	Bushing ring	Clevis
Wheel system	*None*	*None*	*Maintain clearance*
Axle	*None*	*None*	*Transfer load, locate linkage to axle*
E-coat	*Resist corrosion*	*Resist corrosion*	*Resist corrosion*
Label	*Identify assembly*	*Identify assembly*	*Identify assembly*
Molded bushing	*None*	*Retain molded bushing, resist impact, transfer force*	*None*
Jounce limit	*Resist impact*	*Resist impact*	*Resist impact*
Rebound limit	*Limit motion*	*Limit motion*	*Limit motion*
Stone impingement	*Resist stone damage*	*Resist stone damage*	*Resist stone damage*
~~Salt water~~	~~*None*~~	~~*None*~~	~~*None*~~

Figure 4.7 Completed interface matrix for the weldment example. Note deletion of "salt water" as a noise factor in the external section of the matrix.

Then, ask a simple question for each cell: Is there a relationship between the column heading and row heading? If the answer is yes, place a check mark or X in the cell. Do this for all of the cells and save this matrix. When you are first trying to do this, you will probably have difficulty learning to use the active verb–measureable noun construction, and that will effectively interfere with your ability to work through the matrix—and, in particular, to keep the direct, serial relationship rule in place.

After this is done, you can then copy the saved matrix and work through each cell to develop active verb–measurable noun statements at a more appropriate pace. This is also useful if you have a very large matrix. As interface matrices become larger, it naturally takes more time to complete them. Some people (in fact, most people) think they need to complete the entire exercise in one sitting. That becomes impractical for large scopes, so the effort becomes unwieldy. Frustration sets in, and the results often turn out poorly.

If you have a terrible time with the active verb–measureable noun construction—and many engineers do—you can replace the check mark with whatever language you want to use. You can say "the bushing ring needs to be lined up with the clevis so that the axes are OK," which isn't a particularly good statement of function but may be clear to you. You still do need to work through this in a second step, because the active verb–measureable noun construction is critical to making function statements useful in deducing and untangling cause–mode–effect chains.

Finally, using either the check-the-box or write-whatever-you-can-write approach as an intermediate step might be useful if you are unsure about the block diagram and P-diagram results. You might try to construct the matrix, get about halfway through, and decide you need to revisit the earlier diagrams. In fact, this often happens, particularly when you are learning these techniques for the first time.

Overall, though, if you've done a good job with the block diagram and you know the product and usage well enough to construct a sound P-diagram (or at least develop a list of noise factors in the three critical categories), you should have a sound matrix. If you do, and then you carefully work through the interactions, you are very likely to develop a solid set of function statements.

If you've done this appropriately, you should now have a comprehensive and reasonably complete set of function statements. Getting the matrix setup correct and then carefully discussing each cell nearly guarantees a solid result.[8]

CREATING BOUNDARY DIAGRAMS

Another variation that many people use to understand robustness factors is called a boundary diagram—something that is very popular in some automotive firms. (Don't confuse a boundary diagram with a block diagram; that's a relatively common error.) A boundary diagram is usually presented as a tool for discovering interactions between product elements and environmental factors. By creating a boundary diagram, you can make a list of the factors that will address robustness in a design. However, creating a

---

8. There is a technique for assessing whether the list of functions derived from an interface matrix is truly comprehensive. This requires the creation of a function diagram. However, creating a function diagram is beyond the scope of this discussion.

graphical presentation using inductive, imagination-based techniques is far less powerful than doing this same thing deductively.

In fact, a boundary diagram is really nothing more than a graphical presentation of an interface matrix.[9]

If you understand an interface matrix, you'll see that you really don't need a boundary diagram. Moreover, if you are working on a complex system, you'll recognize that a boundary diagram is so complicated that almost no one save the person who created it will ever really understand it. On the other hand, almost anyone can read and understand an interface matrix, even if there are two dozen rows and a dozen columns in the matrix.

The goal of a boundary diagram is to show the interactions between system elements in a visual format. Like interface matrix functions, these interactions should be based on physical interactions, energy transfer, information exchange, or material exchange.

The center of the diagram contains a large box that surrounds smaller boxes representing each element of the design. Items outside the center are other things—environmental issues, usage factors, and other system parts—that define robustness.

Lines connect all boxes or items in the diagram, showing that there is an interaction between any two items. When a system is simple, it's relatively simple to construct a boundary diagram. When a project has a dozen or so elements and perhaps twenty robustness factors, it can literally take days to draw a coherent diagram.

Figure 4.8 shows what a boundary diagram for the weldment looks like—simple and clean, but not as useful as the interface matrix. If you look at this for a moment, you'll see that it hasn't told you anything that you didn't already learn from constructing and completing the interface matrix. And, if you attempted to build this boundary diagram without completing the interface matrix, you'd be much more likely to miss something. Finally, a boundary diagram alone wouldn't give you all of the details about the interactions that detailed function statements provide.

So, when you add up everything, why construct a boundary diagram? The most persuasive reason is to have a graphical tool to show design interfaces—once these interfaces have been deductively assessed using an interface matrix. In other words, it's a good tool for show-and-tell sessions, particularly with managers or executives who are unlikely to take the time to work through the details of reading an interface matrix.

However, I would not recommend building a boundary diagram as an alternative to an interface matrix because it doesn't delineate function

---

9. At least an interface matrix created by the technique just shown.

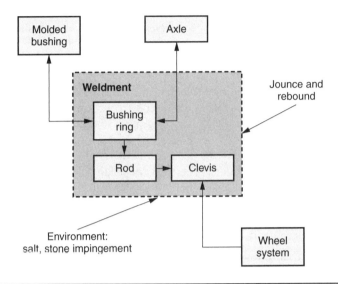

Figure 4.8   Boundary diagram for suspension trailing arm weldment.

statements. And, for a more complicated design than shown above, finding all of the interfaces is difficult, and reading the resulting "spaghetti diagram" with perhaps two dozen or more connecting lines can be nearly impossible.

SPECIFICATION DEVELOPMENT

Once you are satisfied with the interface matrix—meaning you have a good set of rows and columns and you fully understand what interactions arise in each cell—you are ready to get the first major benefit from function analysis.

With the functions extracted from the interface matrix, you should now attempt to create a comprehensive list of product specifications. "Comprehensive" is the key word. Too often, "product specs" are just tables of information that have been handed down, generation after generation of design, and modified slightly for each successive design.

Some are meaningless; others are written in such a way that no one but the original author really can say what the intention of the specification might be. Others are obsolete or described in ways that can only be addressed by physical tests. And, while physical tests are important, modern design—design work that is both effective and efficient—needs to apply more analytical tools and have less reliance on empirical decisions.

In my thirty-plus years of design engineering experience, I've found that really serious problems almost always arise from some aspect of the design that either wasn't addressed in the design information (drawings, specifications, and associated documents) or wasn't properly quantified. Serious design deficiencies rarely crop up from something you've known about for many years.

So, now that you have a reasonably comprehensive assessment of functions from the interface matrix, you need to start a list of the functional specifications for each function that you identified in the matrix. If you have three function statements in a given cell, then three individual specifications are necessary.

For each function statement, you need to decide what limits you wish to apply to that function. These can be attribute statements ("must be green") or variables statements, that is, detailed mathematical accounts ("clevis must not be closer than 2.3 mm in all wheel and axle positions").

Attribute statements can present difficulties. When less explicit issues are involved, you may find that you need to describe a comparison examination, citing reference standards. In other cases, you may be forced to accept general specification declarations—but you should *try* and be as specific as possible. Sensory functions such as "enhance appearance" or "control color" may require a more elaborate definition, such as a jury evaluation, colorimeter measurement, or a visual comparison to a standard.[10]

With variables statements, things are usually easier. You may require one-sided or unilateral tolerances (not to exceed some value, X picofarads minimum, or Y newtons maximum) or you may require two-sided or bilateral tolerances (max/min values, X mm plus or minus 0.1 mm).

To do this, each statement of function requires a specific delineation or quantification of function—how much clearance is needed between X and Y? Would this be a two-sided tolerance, with a maximum and minimum value, or a minimum only? If a beam is to transfer a load, is there a maximum load that must be transferred—a unilateral tolerance? If there is a fixed range, then there might be a bilateral tolerance for this function.

If you've completed the function analysis early in the design process, you may not be certain what values you should select for many or even most of your function statements. Nevertheless, you should at least frame

---

10. However, color can be called out in a high level of quantified detail by specifying chroma, hue, and other colorimetric elements.

the statement, using dummy values or inserting a "TBD" notation to designate that the value will need to be determined later. In other words, you should establish as many preliminary or tentative specifications as possible and be keenly aware that you need to supplant all TBD notations before you release a design.

Doing this provides two important benefits for the remainder of the design process. First, this work has effectively "jump-started" any DFMEA studies that might be done. In effect, the first two columns of the DFMEA have now been completed, and that baseline is perhaps the most important aspect of making a DFMEA worth the work it takes to complete.

Second, this list—or "specification worksheet"—provides a critical checklist for the product designer. If every entry on the list isn't added to the drawing or engineering document set and specific values determined for each variables data function (and less precise statements for all attribute data functions), then you really can't say that the design is complete.

5

Step 3: Deduce Failure Modes

Once you have a clear set of function statements and preliminary specifications for each function, you will be able to deduce a comprehensive set of failure modes (see Figure 5.1). Moreover, you are likely to find that this is far easier and takes less time than you would have believed possible.

To accomplish this, you need to first understand what a "mode" statement really is—and then see how a simple form can be used to thoroughly examine a list of functions for potential failure modes.

To do this properly, you should use a rational, deductive method to determine a reasonable list of failure modes. You should *not* brainstorm for failure modes or use any other inductive, imagination-based method. In fact, once you learn how to do this correctly, you'll realize that if you're working too hard at deducing failure modes, you are probably doing it incorrectly.

FOUNDATIONAL ASSUMPTIONS

Whenever I read a DFMEA and find manufacturing-driven issues in the cause and control columns, I know there's something wrong. The kinds of things I've seen that reflect this problem include:

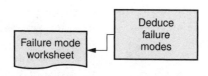

Figure 5.1 Step 3 of DFMEA process.

- Causes that are strictly manufacturing in nature, such as "missed operation," "poor solder joint," "burr not removed," or the worst of all, "operator error."

- Controls that are intended to be carried out when a part is produced, such as "check at assembly" or even good ideas like "mistake-proof operation A."

This could be a result of something simple, such as an inexperienced engineer, but it might be something far more difficult to overcome. This can include a design engineer who's convinced that he's a genius and never makes mistakes. More likely, I've found that these kinds of issues reflect a deep-seated set of conflicts and/or lack of communication between the design engineering group and the rest of an organization, particularly the manufacturing group. If you've got these problems in your organization, DFMEA can't fix this, although it can help in mediating some of the disputes.

Sometimes it's just widespread ignorance about the difference between design verification and production validation that drives these errors.

No matter—the key is starting with constructive failure mode statements. To do this, you must apply two critical assumptions without fail throughout the remainder of the DFMEA process:

1. All hardware and parts (or code entries for software) will be manufactured and assembled according to all released engineering drawings, specifications, and controlling design documents—the design record.

2. All purchased materials, components, and subcontracted manufacturing services will be carried out according to all released product engineering documents and records and supporting purchase orders.

In all honesty, the probability of these two assumptions being correct at all times and under all circumstances is essentially zero. However, the goal is to determine what is *potentially wrong with the design, not* to speculate about what might go astray in producing the part from the design.

In other words, you must assume that the manufacturing community, including all supply chain participants, will execute the design that you release—faithfully in every way—and that any problems that result will be due to errors, including errors of omission, that are made in the design record.

If errors are made in producing the design according to the design record, then those kinds of errors will be addressed using PFMEA techniques.[1] You simply need to assume that once you have released the design record, everything will be done according to the documents in the record.

Any breakdown in applying these assumptions will lead to poor, confusing, or even unusable DFMEA results. This is *extremely* important; far too many DFMEA studies are useless because these assumptions are not followed.

An inability to sort through the difference between design-driven risk and manufacturing-driven risk is a major liability. Once problems arise, both "sides"—and there are nearly always natural divisions in most manufacturing companies—develop a degree of mistrust in the other group.

DEFINITION OF "FAILURE MODE"

Most of the "failure mode" entries in DFMEA studies that have been done over the years aren't really mode statements at all. Instead, they are descriptions of what happens when something goes astray. This usually includes effects and, to a lesser extent, causes.

If you confuse a mode with an effect or a cause, you'll have trouble determining and evaluating risk. Worse, you won't be able to tell the difference between prevention and detection controls, and the resulting DFMEA won't be very useful.

Overcoming this lapse is simple as long as you can resolutely adhere to a simple definition of what a failure mode really is. And please bear in mind: most of what has been written about failure modes, including a good deal of what is written in academic circles and is available on many Web sites, is simply incorrect. If you do a Web search on "failure mode," you will find thousands of pages that describe things that are not proper failure mode statements.

So, we must again work from an operational definition, a definition that identifies clear conditions or events and then tells the user how to state those conditions.

With this in mind, here is the definition we need to employ for a sound DFMEA:

A failure mode is a brief but concise description of how a product may potentially fail to perform or execute a required function.

---

1. If supplier errors creep in due to poorly constructed purchase orders—but not because of errors in the design record—that's another problem that must be addressed. But it can't be resolved in the DFMEA and you should *not* include these issues in your thinking.

Good failure mode descriptions are developed from a customer point of view—and customers include end users or consumers, regulators and government oversight bodies, assembly and manufacturing operations, including transport activities, and all supply chain manufacturing. In this context, "customers" are anyone downstream in the manufacturing-and-sales chain that is driven by the design record.

Just as we learned "column by column, column by column, column by column" in Chapter 4 (Step 2), you need to master another mantra: "fail to perform a function, fail to perform a function, fail to perform a function . . . ."

This definition means that failure modes are simply *negative statements of function,* nothing more and nothing less. Don't go into detail, don't tell a story, don't describe the consequences; just restate each function in a negative way. And, because many functions have more than one possible mode, you need to include every mode that is reasonably possible.

You can understand this better by studying a simple example. Consider a simple, recessed multi-tube fluorescent lighting fixture, the likes of which are a staple in dropped office ceilings all over the world.

What is the function of this product? Well, some people might say, "provide light," but, of course, we know that's a nerd verb and we need something better. So, let's restate that by using the word *light* as a verb and say, "illuminate workspace." This is a measurable function: lumens per cubic meter or lumens per square foot of floor space—depending on the usage factors for the light.

Again, if I ask a group of engineers to quickly tell me the failure modes for a recessed fluorescent fixture, I'd likely get answers like these:

- "Bulb burns out." (That's not a failure mode; that's a cause.)

- "Workspace is too dark." (That's not a failure mode, either; it's an effect.)

- "Light flickers." (Again, this is an effect and not a failure mode.)

These are *not* negative statements of function—instead, they are an attempt to describe *what went wrong,* which will usually lead to a statement describing an effect or a cause rather than a failure mode.

Instead, you could describe the modes for the fixture as follows:

- "Workspace not illuminated." (This is an absence of function.)

- "Workspace incompletely illuminated." (This is a partial execution of function.)

- "Workspace erratically illuminated." (This is an intermittent execution of function.)

- "Workspace illuminated too late." (It wouldn't be possible to illuminate the workspace too early, unless the wall switch was included in the column segment under consideration in the block diagram and scope assessment. Since this isn't reasonable, it doesn't make sense to include a "too early" mode.)

- "Illumination lost over time." (This is a decay of the function under consideration.)

In each instance, the statement of function has simply been altered into a negative statement—no explanation, no fine points. A better way to think about a failure mode is to describe it as a "malfunction" or a "bad" function. This could include the cessation of a function, an interruption of a function, or some other disruption.

In fact, failure modes occur at an instant in time—the function is successful one minute and then it is no longer meeting the measureable and specified requirements. Once a mode occurs, you will only see an effect, or perhaps try to guess about a cause.

So, one of the first things you need to recognize is that any kind of sensory description is *not a mode*—it's usually an effect, but might be a cause or even a control. The thing that trips people up about this, particularly engineers, is that they think too much. If you are one of those people that tend to visualize a scenario in some detail, you will have a strong tendency to state effects when you should be restating function in a negative way.

I know that was true for me at one point, but repeated practice has made it easy for me to mull over an active verb–measurable noun statement and think about all of the ways I can state it in a negative way.

DEVELOPING FAILURE MODE DESCRIPTIONS: THE FAILURE MODE WORKSHEET

This can be done in a systematic, step-by-step procedure once you realize that there are only so many ways that an active verb–measureable noun combination can be restated negatively. The most common ways to do this probably cover 99% of all reasonable DFMEA failure mode statements and include these eight possibilities:

- *Absence of function.* The function is absent from the design. This isn't very common in DFMEA, although it is quite common in PFMEA, particularly for operator-dominant processes.

- *Incomplete function or partial function.* "Partial" is usually the same as "incomplete," but the two ideas may be separate in a few cases. For example, a fluorescent fixture that is always on but distributes light in a spotty manner may "partially illuminate" an area. On the other hand, a fixture that is always on but illuminates one side of an area better than the other would be better described by saying "incompletely illuminates" an area.

- *Excessive function.* It is possible to have too much of a good thing—a fluorescent fixture that excessively illuminates an area is quite unpleasant.

- *Decayed function.* This offers a way to indicate a durability concern, particularly a gradual decline in functionality, in describing a failure mode.

- *Function occurs too soon.* This can only be possible when the function has some sort of sequential role in the overall performance of a product.

- *Function occurs too late.* Again, if the function under consideration is a sequential aspect of a product, this may be possible.

- *Incorrect function.* This is really a last-resort, "catch-all" kind of mode construction. When you know something can go wrong with a function and none of the other categories apply, you can always describe the fluorescent fixture as "incorrectly illuminates" an area—as might be possible if the spectrum of the emitted light was in someway unacceptable or unpleasant. You could, of course, try and explain this in a more elaborate mode statement, but when most people try to create a mode statement with this much detail, they end up describing an effect or a cause.

In my experience, a great way to work through these issues is to use a simple worksheet. What would this worksheet look like?

Of course, it should have some project identification information at the top, and then it should have two columns for information derived in Step 2 of the DFMEA process: a column for function statements and a column for specification information.

Next, a rather large column—the heart of the worksheet—should be provided for the failure mode description. Then, eight "check the box" columns, or *mode-reminder columns,* should follow, with one column for each of the eight possible mode negation conditions.

When you total this up, the resulting form will look something like Figure 5.2.

You'll notice that there are separate row lines for each mode description, but that these lines are absent from the function description section. That's because each function description may very well have more than one failure mode.

Indeed, every function from the interface matrix must have at least one failure mode. I've never seen a design with a function that was so perfect that there was no possibility that the design could go wrong in some way. Some functions, particularly those that have go/no-go or attribute specifications, may well have only one mode.

Most functions have two or three reasonable failure modes—and some have as many as four or five. Over the years, I found that, on average, a typical function will have between two and three modes. If you seem to have more than three modes for most functions, it may well be that the original scope statement and block diagrams are too crude and lack sufficient detail. You may have to go all the way back to that point and review what you've done because things will get very complicated in the next steps if you haven't done a good job in the block diagrams.

The critical factor in determining how many modes apply to each function is the test of reasonableness. Anyone can dream up highly unlikely and even absurd scenarios for a given function that could lead to eight mode statements for a single function (or even nine, if you split partial and incomplete). But would that really add anything to your analysis other than make it into a tedious, plodding, and hair-splitting exercise?

For example, if you were considering a function like "maintain clearance between X and Y," would it be reasonable to include "did not maintain clearance" as a mode? This could only happen in a DFMEA study if there were no dimensions that established a clearance between these two features. If that's the sort of thing that never happens—and hasn't happened in recent memory—it will probably be a waste of time to include that mode in your study.

This means you must exercise a bit of judgment at this point and be a bit less deductive than this form may suggest. However, by working in a team, you will find that these judgments are better and more likely to be practical; you are less likely to become obsessive about including every

FMEA: Failure Mode Worksheet

Project: _____

Page _____ of _____

Date _____

Function description		Failure mode description	Absence of function?	Incomplete or partial function?	Intermittent function?	Excess function?	Decayed function?	Function occurs too soon?	Function occurs too late?	Incorrect function?
Active verb– measurable noun statement	Functional specifications: bilateral or unilateral tolerance/ required properties									

Figure 5.2   Failure mode description development worksheet.

possible mode statement, particularly when the mode is either extremely unlikely or really is a restatement of another mode.

This may require a bit of consideration of causes, but don't get too deep into assessing causal factors at this point. If your team can't seem to agree on mode statements, you might need to discuss causes sufficiently to understand whether a mode is or isn't reasonable. However, don't waste too much time on this. If there's real doubt, not just ego-building semantic argument at work, enter both modes and move to the next function statement.

That's a safe bet because the overall DFMEA process will be self-correcting in this regard. Let's say, for example, that you included both a "partial" and an "incomplete" mode statement at this point in the process. Your team decided to do this after an extended and sincere discussion in which no clear agreement was reached. So, to avoid further delays, you included both mode statements.

Later, when you worked through the remaining steps, you found that the effects and causes were exactly the same, with identical severity and occurrence ratings for each cause and each effect. In other words, the cause–mode–effect chains were equivalent.[2]

Voilá! You've proved that there's really no difference between these two modes.

Looking at the worksheet, though, I believe you can see that developing a list of mode descriptions for a function statement is just a matter of considering each of the eight possibilities—the columns—and, if that's reasonable, writing a mode statement that correlates with that particular possibility. If you put an X in each box as you go, you'll know what you've done. (That's all the other columns are for on this worksheet—just a reminder of what you've done in developing a comprehensive list of failure modes.)

One final comment about this worksheet is in order before we consider an example. This is just a worksheet, and it is not intended to be a reference record. In other words, it is a tool or a "working paper" that you can retain for reference, but it should not be considered part of the audit trail in any quality system or design record system. All of the critical information on this form will be transferred to the main DFMEA form, and that form will constitute the record of your work.

Nevertheless, using this worksheet, or something like it, is a very powerful way to develop failure mode descriptions. And by doing so, you've embraced simplicity while moving closer to a comprehensive understanding of the project.

---

2. Controls are irrelevant in this discussion because controls will be directed at effects and causes—not modes. See Chapter 8 for an in-depth understanding.

LOOKING AT AN EXAMPLE: THE TRAILING ARM WELDMENT

If we again look at the trailing arm example from previous chapters, we can see exactly how the worksheet can be used. In the interface matrix, the internal cell that shows the interaction between the bushing ring and the rod revealed a function of "locate rod to bushing ring." Let's now assume that an approximate and preliminary specification of "14.5 mm from bushing face to rod OD" has been established.

The first possibility is that this function would be absent, or "rod not located to ring." This does not mean that the operation was missed in manufacturing, but instead means that the dimensional requirement was left off the drawing set. So, I would enter that in the "description" column and put an X in the first mode-reminder column.

The next possibility is that the rod would be partially located, meaning that some aspect of the location tolerance, perhaps an angular relationship, details of location tolerances, or some other aspect that hasn't yet been identified in the preliminary specification, had been omitted from the drawing. Again, I would enter an appropriate description and put a check mark in the second mode-reminder column.

The next column is "intermittent function." For the life of me, I can't think of a way that the rod can be intermittently located to the ring. Nor can I conceive of how the rod can be excessively located to the ring. However, I can conceive of several ways that the rod–ring location can change over time—a weld that's too flexible or a rod diameter that lacks strength can lead to this—and so I would put an X in the "decayed function" mode-reminder column.

Finally, I can't think of anything that's design-driven that could cause the ring to be located to the rod too early or too late. However, I can anticipate that it is not impossible that we will include all of the dimensional call-out information for this relationship on the drawing, but that some portion of this information may be incorrect. Therefore, I will add one last mode, namely "incorrectly located," and place an X in the final mode-reminder column.

Then, on further reflection (or better yet, discussion with team members) I realize it's more than a bit absurd to think that we could possibly leave this off the drawing set completely. We've been designing trailing arms for years and we've never forgotten to dimension this relationship before. We have made mistakes by entering a partial call-out of this relationship or made errors in getting the dimensions correct for the complex suspension geometry this part fits into, but we've never forgotten the dimension per se. So, we'll just cross this description off our list.

Overall, you've put one X in one of the columns for each failure mode description that you think may be reasonable. You can even have two mode descriptions—two rows on the form—with the same mode-reminder column checked, as could happen if "partial" and "incomplete" represent different modes.

Once in a while, you may choose to put two X's in a given row—a given description—but that should be rare. You could almost always check the "Incorrect Function" column whenever you've checked another column, but the worksheet will be more useful if you save that column for a distinctly separate mode description that has a different meaning.

So, this small portion of the DFMEA worksheet would end up looking like Figure 5.3, with three mode descriptions after the "absence" mode has been deleted.

Ultimately, the X entries in the mode-reminder columns will tend to have a top-down, left-to-right echelon structure for each function, but the X's really don't matter very much in the overall scheme of things. The only things that really matter on this worksheet are the actual failure mode descriptions.

You would then repeat this process for all of the functions derived using the interface matrix, and at that point you will find that you have the first three columns of the actual DFMEA form completed. And, you will realize that you've been working column by column.

We first established all of the functions using the interface matrix technique. We then created preliminary specifications for each function, and now we have deduced—not brainstormed, not imagined (at least not very much)—all of the reasonable failure modes that devolve from the functions.

When you've completed the failure mode worksheet, you will be able to move all of this information directly to the main DFMEA worksheet. If you're using a computer-driven spreadsheet for this, you can do this quickly. If you're really skilled, you can even build a macro system to transfer all of this without manual cut-and-paste actions.

However, all of that is just icing on the cake. The real goal is to deductively develop a sound list of failure modes on a function-by-function basis.

A FEW GOOD TIPS: GETTING BETTER RESULTS

After you've derived failure modes for several functions, you'll start to notice a pattern. The subset of modes for each kind of function will repeat, over and over again. Most clearance functions in a mechanical design will

FMEA: Failure Mode Worksheet

Project: _____

Page _____ of _____

Date _____

Function description

Active verb– measurable noun statement	Functional specifications: bilateral or unilateral tolerance/ required properties	Failure mode description	Absence of function?	Incomplete or partial function?	Intermittent function?	Excess function?	Decayed function?	Function occurs too soon?	Function occurs too late?	Incorrect function?
Locate rod to bushing ring	14.5 mm from bushing face to rod OD	Not located	X							
		Partially located		X						
		Loses location over time					X			
		Incorrectly located								X

Figure 5.3   Failure mode worksheet with modes for one function.

be the same, for example. And that means the work needed to get a comprehensive list of failure modes won't take much time once you get rolling.

If you consider the single-function example shown in Figure 5.3, the total time to work through the final list of three failure mode descriptions would be less than five minutes for a team without any experience. This could well be less than a minute for an experienced team.

Once you get experience with the technique and have a good worksheet design, you'll find you can work through even a very long list of functions in a relatively short period of time. The only caution, though, is not to get too carried away with cut-and-paste activities. Make sure you take enough time to consider each function and the possible modes in sufficient detail to be confident that a pattern from an earlier, similar kind of function does, in fact, repeat.

So, to sum up, you need to keep a few things in mind to get a good set of failure modes:

1. Use a worksheet approach; even if you are very experienced and have memorized the eight negative possibilities, a worksheet is a great aid.

2. Don't think too much. Try to make this a deductive activity rather than a deep, introspective, and inductive bit of work. Almost everyone struggles with this tactic the first time, particularly if you've done FMEA work before and need to unlearn some bad habits.

3. Try and enter modes that are reasonably possible and not once-in-the-lifetime-of-the-universe events.

4. If you get stuck on what is reasonable, err just a bit on the side of additional modes. You can always take these out later if you find they don't tell you much.

5. Don't let manufacturing issues creep into your thinking. This can be difficult, but keep focused on what can go wrong during the actual design work. What errors can a designer, product engineer, or design engineer make?

6. Always remember that the worksheet is a memory tool, not a high-level record of your work. Quality system audit trails are complicated enough. Don't add to the paper trail when additional records won't tell you much about what's important.

7. Keep the customer viewpoint in mind when thinking about mode descriptions. And remember that customers include regulators,

supply chain partners, shipping firms, dealers, sales people, manufacturing operations, and service personnel.

8. If you start describing what you can see, feel, smell, touch, hear, or taste, you will probably *not* be describing a failure mode. Instead, you'll be describing a failure effect, or perhaps a failure cause. Just restate functions in a negative way and you won't fall into this trap. If you can sense it, it's not really a mode.

9. Don't get stuck on where to put the X's in the mode-reminder columns. Again, the critical information on this worksheet is the list of failure mode descriptions. You should have only one X in each row—with one mode description—most of the time. You'll see an echelon of X's on the right-hand side of the worksheet if you have more than one mode. Work through the mode-reminder columns one mode at a time and you'll see how easy this is to do.

6

Step 4—Effects and Severity

I t's certainly taken some effort, but we are finally ready to start working on a subject that many people who've done DFMEA studies in the past will find more familiar: effects and severity (Figure 6.1). Even in this area, though, there are surprising and powerful ideas that can make DFMEA more worthwhile than you thought possible. And—even more importantly—you will probably find that these ideas will make the process go faster while providing more insight than you've seen when using other methods.

In order to start this step, let's first review what we've done by looking at a small portion of the main worksheet, complete with the information we've developed for the trailing arm weldment in the previous steps.

First, we explored the scope and determined that the weldment had three elements—the rod, the bushing ring, and the clevis. We then examined critical noise factors by creating a portion of a P-diagram and used the block diagram and P-diagram to construct an interface matrix. With this information in hand, we were able to complete the header section of the main DFMEA worksheet or form.

Using the interface matrix, we examined the interactions between weldment elements—and then the interactions between these elements and critical noise factors. For each interaction, we developed function statements consisting of active verbs and measurable nouns.

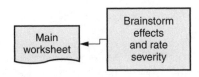

Figure 6.1   Step 4 of DFMEA process.

Item/function	Functional requirement(s)	Potential failure mode
Locate rod to bushing ring	14.5 mm from bushing face to rod OD	Rod partially located
		Rod loses location over time
		Rod incorrectly located

Figure 6.2 The DFMEA worksheet, with information transferred from the failure mode worksheet.

And, for each function statement, we attempted to define a preliminary specification, both demonstrating the measurability of each statement and confirming that we had, in fact, defined a meaningful function.

Then, for each function statement, we deduced failure mode descriptions, always keeping in mind that a mode statement is just a negative statement of function. By the time all of the functions had been assessed for failure modes, we had completed the first three columns of the DFMEA record form.

For the trailing arm weldment, we actually worked through one row of the form—bearing in mind that all of the first three columns on the main worksheet would actually be completed before moving on to Step 4 in the DFMEA process.

So, for the trailing arm weldment, we can now transfer information from the failure mode worksheet to the main DFMEA worksheet, and our single-row example—expanded into three sub-rows, based on three reasonably probable modes—looks like Figure 6.2.

UNDERSTANDING EFFECTS—AND CAUSE-AND-EFFECT RELATIONSHIPS

Just as we did for the term *failure mode,* we need to start with an operational definition for *effect.* For a deductive DFMEA study, here is the most useful operational definition:

A failure effect is a description of the consequences that may arise from the disruption of a function—or, more specifically, from a failure mode.

Each failure mode must have at least one effect, but most failure modes will have more than one reasonably possible effect.

Ultimately, effects must crop up from causes, and this means that we have entered the dangerous territory of cause-and-effect analysis. The idea of cause and effect is deeply embedded in modern human society. This can be very simple or extraordinarily complex, depending on what you are looking at. However, most engineers are quite sure that they understand cause-and-effect relationships—but the overwhelming experience that comes out of modern product development systems shows that this understanding is sometimes tenuous as best.

Most engineers believe that a specific cause will lead to a specific effect. For example, if I throw a rock against a window, the window may very well break. A cause—a thrown rock—results in an effect—a broken window. The broken window doesn't cause the ball to be thrown. Is it always that simple?

In the real world of product development, reversing or confusing the relationship between causes and effects is not as simple as the rock-into-the-window example, and this has led to many product deficiencies over the years, sometimes with serious consequences. Moreover, this has been—and continues to be—a major problem in developing new technology. Furthermore, confusion about cause-and-effect relationships often causes DFMEA studies to become dreary, time-wasting exercises.

Anyone who has worked on DFMEA studies and has used a row-by-row approach rather than a column-by-column method has struggled with this. I've actually observed teams spend an hour or more debating what constitutes a cause and what represents an effect for a single failure mode. Once this kind of confusion starts, it seems that causes are inevitably confused with effects.

If you follow the methodology explained in this book, you will have to make a conscious effort to make this error. We've already started out correctly, by developing well-structured function statements and deductive failure mode descriptions. Now we need to examine exactly what it means to say that a "cause-and-effect relationship exists."

Let's start by trying to forget all of the deductive work that we've done on the weldment assembly and also forget about the "column by column" concept. If we started by brainstorming failure modes (rather than deducing the modes) for the weldment, it's very likely that we'd get a mode like "car rides rough."

Of course, this isn't a mode. It's a sensory event, so it must be a cause or an effect. Now, let's further assume that someone, somewhere realizes this and actually works through the function–specification–mode process and determines that a proper mode is "location of bushing ring to rod is lost over time."

Is "rough ride" a cause or an effect? You can almost hear the debate.

- A rough ride causes either a plastic or fatigue-driven deflection of the weldment; therefore "rough ride" is a cause.

- No! A rough ride is a result of the gradual loss of relationship between the rod and ring. That means it's an effect.

This kind of debate can go on for a long time. However, in this case, one side of the debate provides a useful answer, while the other side provides an answer that's less useful. (We'll let the angels debate who is "right" and who is "wrong." These kinds of arguments are really counterproductive in DFMEA because we are looking for insight and better designs, not abstract concepts like truth . . . wise men like Diogenes can discuss these issues.)

How can we know what the most useful approach might be? To explore this, we need to take a deeper look at cause-and-effect relationships. And, in doing so, we'll see that the conventional wisdom, typically outlined in a fishbone or Ishikawa diagram, doesn't serve us well. That doesn't mean that a fishbone diagram is incorrect; instead, we will find that we can learn more from recognizing that the underlying logic of a fishbone diagram isn't as comprehensive as it needs to be to look at cause-and-effect relationships in DFMEA.

Once a product is created, there are a number of functions that the product can accomplish. At the same time, there are many causes that are "pushing" against each function. Diagrammatically, this looks like Figure 6.3 for each function.

Now, if any of these causes has sufficient impact, the function might be disrupted. This will result in an error state or failure mode, as shown in Figure 6.4.

Once the function is disrupted, there may be any number of consequences—or effects—that might ensue. This completes the cause–mode–effect chain and the resulting diagram now looks like Figure 6.5.

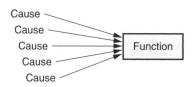

Figure 6.3 Causes "pushing" against each product function.

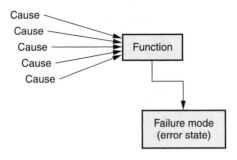

Figure 6.4 Function disrupted by causes, resulting in failure mode or error state.

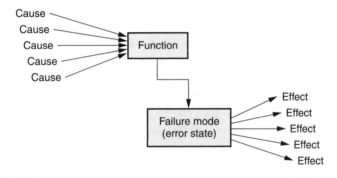

Figure 6.5 Effects due to failure mode caused by disruption of function.

This looks a bit like a traditional fishbone diagram, but, in fact, it reveals much more than a typical cause–effect analysis would typically divulge.

First, if you look closely, you don't need to know anything at all about the failure mode or the effect to speculate about causal factors. What could cause the weldment to lose the relationship between the bushing ring and the rod? We don't need to know right now, because we are concerned about the effects, and the effects are things that occur after the function under consideration goes awry.

In sum, causes are actions, events, or decisions that come to the fore *before* an error state arising or a disruption of function. In other words, any causal factor or "cause" will be something that occurs before any appearance of the error state. And, in DFMEA, that is something that is done in the design process, not in manufacture or use of the product or service.

Furthermore, we don't need to know a thing about causes to speculate on effects. Worrying about this will just cause confusion for most people. For our purposes, we simply need to consider all of the things that might arise *after* the error state commences. Anything that takes place once a function is disrupted will be an *effect* and *not* a cause.

More to the point, we don't need to understand which cause leads to which effect. In DFMEA, that's a very significant yet subtle idea. In essence, we really don't know and don't need to know what specific cause leads to which effect—even though most people believe this is a critical issue.

If you think about this diagram, you will see that the situation described is incredibly complicated. How many cause–mode–effect (C-M-E) chains are possible if there are five causes and five effects?

The answer may surprise you, but the number of possible C-M-E chains is staggering. A complete inventory of cause–mode–effect chains would consist of:

$$\text{Number of C-M-E chains} = N_C! \times N_E!$$

In other words, the sum of all possible chains would be the product of the number of effects factorial and the number of causes factorial.[1]

This can occur because one effect may arise from one cause, or from two, three, four, or even more causes. Similarly, two effects could occur as a result of one, two, three, four, or five causes. If you tabulate all of the possible combinations, the total is huge.

If there are five possible causes and five possible effects, this means there are $5! \times 5!$ combinations of cause and effect, or 120×120 or 14,400 possible combinations. Analyzing all of these potential combinations would require a long time, perhaps a lifetime—or at least the career of one engineer.[2]

DON'T LET COMPLEXITY OVERWHELM ANALYSIS

How can we cope with this level of prospective complexity? Since we really do need to brainstorm to get a sound understanding of effects, we nonethe-

---

1. A number factorial means to multiply that number by all the whole numbers below it—4 factorial would be $1 \times 2 \times 3 \times 4 = 24$ and 5 factorial would be $1 \times 2 \times 3 \times 4 \times 5 = 120$.

2. Worse, at least 90 percent of these scenarios would be so unlikely as to be bizarre—and DFMEA teams often spend time debating just such scenarios.

less need to keep a check on the whole business of conceiving likely or reasonably possible effects.

All we really need to understand—at least at this point—is what can happen after the function in question is disrupted.

Moreover, the most important need is to identify the most *significant* effects. Don't worry about causes; don't worry about the complexity of C-M-E chains—just consider the most serious effects. And always remember that effects occur after the mode has emerged, or more appropriately, *after* the function has been disrupted.

In the weldment example, let's look at one possible cause–mode–effect chain. At this point, we only have functions and modes, but that's the starting point. For example, we know that the function of "locate rod to bushing ring (14.5 mm from bushing face to rod OD)" could be disrupted by "losing location of rod over time."

What can happen once the rod and bushing ring are no longer correctly located? Here are some possibilities:

- Rear tires become misaligned

- Tires wear prematurely

- Steering precision is diminished

- Vehicle ride is degraded

- Rear wheel traction is decreased

In fact, I'm sure that most chassis engineers could add to this list—but I'm also confident that this is a fairly good list. And, we didn't need to know or speculate in any way about what might cause these effects to arise.

How many of these effects need to be used in the DFMEA study? To start with, it is absolutely essential that the most *severe* effect be entered in the form, and then used in the remainder of the study. You can certainly enter more, but at some point you will find that little is gained by doing so.

In practice, I've found that listing more than two effects—and, in most cases, listing more than one effect—needlessly complicates any DFMEA study.

Because this is a controversial subject in the DFMEA and quality communities, it's important to explain why entering all effects is usually an unproductive exercise. To do this, we need to stop and think about why we are conducting a DFMEA. There are really three major reasons:

1. DFMEA helps you identify and address weaknesses in design.

2. DFMEA provides powerful guidance for more comprehensive design verification, reducing the probability of release of designs with avoidable failings.

3. DFMEA allows design teams and management to assess risk and decide how much investment and product cost will be allocated to specific issues in any design.

Ultimately, this means cause–mode–effect chains that present unacceptable risks[3] must be dealt with in one way or another.

What can you do?

If you think about the diagrams presented above, you'll start to understand that the most powerful and useful way to reduce risk is to address the causes. In all but a very few cases, you can do little or nothing to impact what effect might occur once a function is disrupted.

If the rod and bushing ring are no longer properly located, can the design of the weldment really have any meaningful impact on which of the effects listed above may occur? This is a more complicated web of probabilities that really requires fault tree analysis (FTA) to untangle— and a sound FTA study makes DFMEA look like child's play. This is true because we don't estimate the occurrence of effects in FMEA; we estimate the occurrence of causes. But that's something we need to deal with in Chapter 7.[4]

Another example from the automotive world will help illustrate this idea. For a seat belt, one of the important functions is "maintain latched position." Among the reasonably possible failure modes will be "partially maintains latched position."

If this disruption occurs, there are several things that can happen:

• Latch vibrates and makes noise

• Latch entangles clothing

• Latch opens during impact event

These effects are not equally probable, nor are they equally serious. In fact, the most probable effect is that the latch will jangle in some annoying way. But this is far from the most serious; the possibility that the latch might open during an impact is much more critical, but much less likely. Getting

---

3. We'll look at this in more detail in Chapter 9.

4. This is a very subtle idea that most people hear and think they understand. However, in most cases, the subtlety is completely lost. You need to "roll this around" in your head for a few moments to really comprehend what this implies.

your clothing caught in the latch is less serious than the latch opening during a crash, but less likely than the belt mechanism rattling because it is partially latched.

So, what can you do about any of these effects? In terms of the effect, the answer is not much. Once the latch is partially engaged—regardless of the cause—you can't be sure what will happen, only that several effects are possible. This means that, as a design engineer, you really can't control all of the events that might arise if the latch is partially engaged.

In this case, I think I'd be much more concerned about the latch opening in a crash than I would be about a rattle in the mechanism. That doesn't mean that the rattle is acceptable; instead it suggests that we are more likely to focus attention on any C-M-E chain when the seriousness of the latch opening in a crash is considered.

The reality is, though, that you can only do something about this by addressing the causal factors that could lead to a released design in which the potential for a partially engaged latch is higher than some relatively improbable level, say six sigma, for example. We'll continue to expand this discussion in the rest of the book, but for the moment, we need to recognize that addressing causal factors will reduce the probability that a given mode might arise.

Properly addressing causal factors can and usually will lead to this chain of events:

- The probability of the occurrence of some specific cause (a cause of a given mode, we don't need to speculate about what effect might come up) will decrease.

- Therefore, the probability that a given function will be disrupted will decrease, since the cause is less likely to come to pass—and the disruption will therefore be less likely.

- And, as a result, the conditional probability that a given effect will crop up is decreased—not because any action was taken to inhibit the effect, but instead because action was taken to prevent the mode from arising in the first place.

When you dig into the full range of conditional probabilities—the 14,400 possibilities that five causes and five effects for a single mode can exhibit, for example—you can quickly become confused. Just discussing this can make most engineers a bit queasy. It makes my head spin at times and I've been working in this area for more than thirty years.

The easiest thing to do is to remember a single, simple rule: *Effects occur after modes arise.* Period. You don't need to think about this as much as you want to or are used to. This isn't quite as mechanical as deducing modes, where excessive thinking is a real drawback, but you still don't want to try and make your analysis more complicated than DFMEA requires.

And, in DFMEA, effects are things that are *not* design decisions. In the example cited earlier, a "rough ride" happens after the rod–bushing relationship is changed—it is therefore an effect. And it can't be a cause because "rough ride" is not a design decision.

Finally, if you want to enter every possible effect you can think of on the DFMEA, you are certainly welcome to do so. If you are working on a completely new design, without any previous history to guide you, you may want to enter more than one effect. However, as we will see, adding these effects does little to help you understand the risk associated with C-M-E chains—and does almost nothing to help you address risk.[5]

As a general rule, you *must* enter the most perilous effect. (We'll see how to determine which effect is the riskiest once we tackle the subject of severity.) If you want to enter a second, feel free to do so. However, you won't gain much in most instances, and entering more than two effects rarely has any practical impact on a DFMEA study.

If you insist on entering every effect—and, in the next process step, every causal factor—you'll be addressing thousands of C-M-E chains before you know it. You won't learn much about your project by doing this, and soon you'll start to think that DFMEA is a terrible waste of time.

"CUSTOMER DISSATISFIED" IS NOT AN ACCEPTABLE EFFECT

This is one of the worst errors you can make in FMEA studies. You should never, ever use the phrase "customer dissatisfied" to describe what occurs after a function is disrupted. If something goes wrong, it is certainly true that a customer is very likely to be dissatisfied. However, this simple-minded statement, even though correct, tells us nothing at all about the consequences of a failure.

If a seat belt latch rattles, I will almost certainly be dissatisfied. If it opens in a crash, I might be a wee bit more dissatisfied. Well, maybe I wouldn't be around to be dissatisfied in a serious crash, but I hope that my wife would be . . . well, maybe not! (That's a chain of conditional probabilities that I don't care to untangle.)

---

5. On the other hand, the only thing you "must" do is pay taxes and die; if you want to enter ten effects for every failure mode, you can do so. It just won't add much to your DFMEA study.

The problem with a statement like "customer dissatisfied" as an effect is that this doesn't tell you much about *why* a customer might be dissatisfied. Because you don't know the depth of this reaction, you won't be able to rate the severity of an effect properly.

So, why does this show up in so many FMEA worksheets? In my experience, the most common reason is poor technique—such as using row-by-row methods—that confuses modes and effects. If you use poor technique, you may not deduce modes that are negative statements of function. Instead, you are likely to enter an effect in the mode column on the worksheet.

Once you confuse mode and effect, you are then left to ask, "What is the effect of this effect I've called a mode?" And it's all too easy to see that the only thing left to say is that the customer will be dissatisfied.

If you find that you want to enter "customer dissatisfied" in the effect column, go back and review your work on functions and failure modes. You're very likely to find that you've done something wrong at these earlier points in the process.

ANOTHER PITFALL: TOO MUCH "WHAT IF" THINKING

Another possible problem that can appear when you are brainstorming effects is to let your imagination run wild. You should only try and imagine the *most reasonable and proximate effects,* using the perspective of the customer. In fact, in the same way that almost all humans are connected by six degrees of separation, you can easily find that asking "what if" six times can lead to a prediction that the end of civilization will result for almost any failure mode.

For the weldment, let's assume that the loss of rear wheel traction is the most severe effect that can come from the loss of location of the rod to the bushing ring over time. Of course, a loss of traction could lead to an accident. And what if that accident occurred in New York near the United Nations complex? And what if your vehicle then collided with the limousine carrying a head of state? And then what if that leader was killed, and his nation concluded that your act was a deliberate attempt at terrorism? And then they decided to attack the U.S. capitol, setting off a worldwide nuclear exchange?

Of course, that's an absolutely crazy scenario. But it could happen—and I've heard dozens of somewhat-less-likely scenarios argued in DFMEA groups over the years. In the end, all of these kinds of "what if" discussions are a terrible waste of time. They add nothing to your understanding of a

design or project and, in reality, they are another form of conditional probability that design engineers have no control over and can do nothing about.

In the end-of-the-world situation cited above, the only thing that a
design team can control is the weldment—and one of the effects that can
be addressed is the loss of rear wheel traction. The design team has no control over the use of the vehicle in the vicinity of the UN complex, no control
over the presence of a major head of state, and no control over the reaction
that another nation might have to an accident that comes about because rear
wheel traction is degraded.

Engineers are trained to think in detail, to be precise and complete
in their thinking, and to honestly and forthrightly report their work.
However, this can easily lead to a near obsessive-compulsive need to be
"comprehensive." Falling into this trap, whether it comes about when you
are attempting to deduce modes or imagine effects, is a delusional form of
perfectionism that benefits no one. It does allow some people to show off
their mental agility and capacity to construct imaginative and even entertaining incidents, but the DFMEA process isn't the place to do this.

Just don't do it—stick to realistic and immediate descriptions of effects.
In the end, a good list of effects will tell you one or two of the most serious
things that might happen once a failure mode occurs.

TECHNIQUE CONSIDERATIONS FOR
EFFECT-SEVERITY ANALYSIS

Once you've overcome most of the complexity concerns we've just described,
you are actually ready to start generating information that can be entered
on the main form.

The best way to do this is with a modified brainstorming technique.
Working column by column, consider a single failure mode and brainstorm
the various effects that the DFMEA team thinks are possible.

Write these down, either on a whiteboard, easel, or piece of paper.
Don't try to rate the severity of these effects, but instead start by picking
the one that is the most serious. Enter the most serious effect on the form.
If you think another effect is almost equally serious, you may also want to
enter this on the form.

Stop after choosing one effect—or possibly two if they are equally serious or nearly so—and move to the next mode. If you have a good reason,
you may want to enter more than two effects, but, again, I urge caution.
Most of the time, this is a time-wasting and distracting process that adds
little or nothing to the DFMEA study.

Item/function	Functional requirement(s)	Potential failure mode	Potential effect(s) of failure
Locate rod to bushing ring	14.5 mm from bushing face to rod OD	Rod partially located	Rod–ring interface fractures; loss of vehicle control
		Rod loses location over time	Steering precision is diminished
			Rear wheel traction is decreased
		Rod incorrectly located	Premature tire wear

Figure 6.6 DFMEA worksheet with failure effects.

It should not take a great deal of time to do this. The first few effects may take a few minutes each, but soon a well-oiled team should find that each effect discussion is no more than one or two minutes in length. Occasionally, a more protracted discussion will arise, but this should be an exception.

If you enter more than one effect, you will have to add rows into the form. So a single function may split into two, three, or more modes. Each mode may then split into two or more effects, creating a large number of potentially unique C-M-E chains—chains that really don't help improve the design much and create dubious entries that consume time and energy.

For our example problem, our worksheet could now look like Figure 6.6—bearing in mind that this is just an example, and the actual entries would be different if a team of experts produced it.[6]

UNDERSTANDING AND RATING SEVERITY

The second thing we need to do in Step 4 is to describe the way customers might react once a function is disrupted. The reaction of customers to a failure mode is termed *severity*. Severity is a principal factor in assessing risk, and so we need to have a sound operational definition for this term.

---

6. I've deliberately included a two-effect line so that the branching of C-M-E chains can be seen.

For deductive DFMEA:

Severity is a numerical rating of the impact of an effect on customers.[7] To rate severity, a table is used to create a standard set of criteria that will allow a given effect to be rated and a number assigned to the effect.

A low number means that the effect isn't very severe; it may not even be noticed in some cases. A high number means that the result is serious, and the highest numbers in most tables are reserved for the potential of injury or death.

The construction and use of severity tables can be quite complicated. In most industries, a standard table, such as the table provided by the Automotive Industry Action Group (AIAG) in the automotive industry, is the table that should be used. The AIAG table has a low value of 1 and a maximum value of 10. Historically, severity tables have been based on a 1 to 10 standard, although the specific rating descriptors have varied from situation to situation.

In healthcare, very complex tables with a scale from 1 to 16 have been developed and, in these tables, developed by the Veterans Administration, the frequency of occurrence of the underlying effect is included in the table criteria. While this is an elegant and clever addition to the methodology of FMEA, in my experience it adds a level of unneeded complexity and subsequent confusion about assessing risk. And, in any event, healthcare FMEA studies are nearly always process studies—so most DFMEA practitioners would do well to steer away from these kinds of tables.[8]

In the end, you should use a table that is common in the business and/or industry that you serve. If you work in an area that has no generally accepted tables, I've provided a set of generic tables in the appendix of this

---

7. Notice that the term is "rating," and not "ranking." Even though many sources describe severity as a ranking, ranking means to order several terms or ideas relative to one another, while a rating is a comparison against a standard. In FMEA, we compare against a standard, namely a table; we do not "rank order" effects in some way, and so "ranking" is a completely erroneous term. This is a major "geek level" error in the Automotive Industry Action Group FMEA reference manuals that have been published over the years.

8. These tables also produce some ethical quandaries, in my opinion. In the Veterans Administration tables, a rare but fatal effect presents the same risk as a frequent annoyance that has no long-term impact on many patients. That's a position that would be difficult to explain to the family of someone who's died as a result of a design defect.

book. Feel free to use them as you see fit, but bear in mind the limitations of tables in general.

Because tables are used to generate numbers—and the numbers generated are used to assess risk and prioritize corrective actions—teams tend to spend a great deal of time debating the content and use of tables. Unless you have a high degree of reliability expertise, I strongly recommend that you don't do this. The potential legal implications of using an ad hoc or nonstandard table can be significant, and doing this without the proper expertise and consideration can cause serious difficulties. We'll address this in more detail in Chapter 10.

Keep in mind that tables are never perfect. It's impossible to create a 1-to-10 scale that concisely and accurately differentiates between different levels of risks associated with all possible effects. No matter which table you are using, a value of 1 is something that is almost invisible to a user or customer. A value of 10 is something that presents the highest level of risk, whether that risk is physical or financial. A value of 9 is usually reserved for an effect with the highest level of physical or financial risk—but the effect itself provides some kind of warning to the user or customer before the full impact is felt.

To use tables properly, follow these guidelines:

1. Complete the entire column of effects (work column by column) for all modes before discussing severity in any detail. Then and only then should you move to rating severity by using a table. Early on, a team will have some difficulty in deciding which effects merit a particular rating value.

2. Once you have rated a few effects, the rest of the effects will likely be much easier to rate. In fact, many of the effects are apt to repeat, and so there will be almost no effort at all in rating many of them.

3. Because you are working with just one table, you won't get confused; if you are working row by row, you'll be jumping from table to table to table and things can get messy.

4. If you are working column by column, you won't find that your understanding or application of the table entries "drift" as you move to the end of the DFMEA study. (This is a common event when row-by-row techniques are used—you see similar or even identical effects with different severity ratings, particularly in a long DFMEA study.)

5. Don't think that a table—no matter which table is being used—will have a set of perfect descriptors for every effect in your DFMEA study. That won't occur, and there will always be one or two aspects of the effect that seem to fit with more than one rating value. A severity table is a guideline, not an absolute scale.

6. If you become enmeshed in a debate about which value to assign to an effect, particularly if the debate is about a one-number difference, don't argue. *Just pick the highest value.* You're far better off being a bit pessimistic; you can always change the number later if you find better data that will support a lower value.

7. If a number debate is related to a significant difference (three numbers or more), you really need to discuss and review this in more detail. The difference between a minor annoyance and the possibility of injury is huge, and if your team doesn't see eye-to-eye on something like this, it's entirely possible that one or more team members don't understand how the product or service is used or perhaps how it works.

8. After rating the first half-dozen or so effects, you should be able to rate most effects in a matter of seconds, not minutes or hours. This means that the rating process, after an initial effort by the team to get its bearings, should go very quickly. If you find that your team is taking a long time for every rating, you need to step back and figure out why this is happening.

SUMMING UP: STEP 4

Even though the ideas that drive good analysis, particularly those that underlie cause–mode–effect chains, can be complex and require some serious consideration, the overall effort to complete this step should be considerably less than you needed to develop a good list of functions and corresponding failure modes. After all, that's one of the major benefits that a deductive approach to DFMEA promises, but it won't be automatic.

Keep it simple. Look for the one or two most serious and immediate effects for each mode. Then, after completing the column with all of the modes, get out the severity table and start rating the effects.

The simplicity of this step can be summed up in the diagram shown in Figure 6.7.

Following this diagram, our worksheet for the weldment could look something like Figure 6.8, using the AIAG table set to rate the severities.

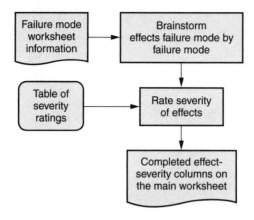

Figure 6.7 Step 4 activities.

Item/function	Functional requirement(s)	Potential failure mode	Potential effect(s) of failure	Severity
Locate rod to bushing ring	14.5 mm from bushing face to rod OD	Rod partially located	Rod–ring interface fractures; loss of vehicle control	10
		Rod loses location over time	Steering precision is diminished	7
			Rear wheel traction is decreased	7
		Rod incorrectly located	Premature tire wear	5

Figure 6.8 The main worksheet with effects and severity ratings.

7

Step 5—Causes and Occurrences

With all of the important effects listed and the severity of these effects entered on the worksheet, we can turn our attention to causes (see Figure 7.1). In this step, we will ask two relatively simple questions:

- What events are responsible for a failure mode occurring?

- How likely is this set of events?

In Step 4, we took a good look at cause-and-effect relationships, and to conclude this, we must start again with an operational definition for "cause." However, the lengthy cause–mode–effect chain discussion in Chapter 6 has actually explained a good deal of the underlying assumptions and concepts needed to understand causes.

With those ideas in mind, here is a definition for "cause" you will find useful for deductive DFMEA studies:

> A failure cause is a description of the fundamental reason, sometimes called a "root cause," *consistent with the project scope,* that gives rise to a system or part failure mode.

Let's deconstruct this a bit at a time so that we can fully understand what this means. First, a *fundamental reason,* or *root cause,* is something that,

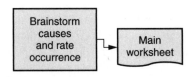

Figure 7.1 Step 5 of DFMEA process.

in some percentage of cases—but rarely always—leads to a disruption of function.

In many or even most instances, there will be more than one root cause. These causes may work in concert or work independently, and this will lead to some complexity in this step. In fact, the term "root cause" may be a bit misleading because there are often several factors—which we can call causal factors—that lead to a failure mode.

Second, a cause that is within the scope of the project is a cause that is *within the control of the design team.* This means that a sound causal factor must be a *design error*—something that is a mistake that is made by a product engineer or designer. A partial, but far from comprehensive, list of DFMEA causes can include:

- Calculated tolerance, size, or shape incorrectly.

- Analyzed circuit improperly.

- A component or part is omitted from a subassembly—as a design error, not as a manufacturing error.

- Used an inaccurate model or simulation.

- Selected the wrong material or specified incorrect material properties.

- Structured a conditional software loop incorrectly.

- Incorrectly entered dimensions—which can include conceptual errors, simple omissions, improper use of dimensional standards, such as geometric dimensioning and tolerancing, or some other drawing error.

- Used incorrect design assumptions—a causal factor that can also include a supply chain customer providing incorrect or inadequate design or product information.

- Improper clearance between elements.

- Excessive friction, wear, or adhesion of some sort.

- Incorrect pressure, voltage, current, or force specified.

The list is endless, because the number of errors that a design team can make is limitless. But the idea is to keep causal factor descriptions within the scope of the project while limiting these descriptions to design errors.

At the same time, it is critical that you not use descriptions that are either outside of the scope of your project or are not design errors. The three most common incorrect DFMEA causes include:

- Manufacturing errors of any kind, including "operator error" in any form.

- Abuse by users or customers—if this appears to be important, it should have been included in the P-diagram and used to generate functions and modes.

- Any kind of error made in the supply chain other than provision of incorrect product information that affects the design.

These kinds of "causes" are, in the end, an acknowledgement of something that we discussed in Chapter 2: design engineers have difficulty thinking about their own errors. In many ways, this is a natural response. A proficient design engineer is knowledgeable about the product being designed and must exhibit a high level of confidence in decisions that he or she will make. That makes it difficult, verging on next to impossible, for anyone who's made decisions about a design to see what mistakes might have been made.

As a result, design engineers tend to see problems as the result of something that is outside of their control. This is a fantasy, of course, but it is a subconscious belief that often has a major impact on design decisions, and DFMEA in particular.

Quite honestly, no competent design engineer ever deliberately makes an error in design. Instead, errors tend to be inadvertent, based on honest wrong beliefs, or some other aspect of complexity that arises in the development and execution of the overall project. So, make sure that you stay away from trying to find some source of error outside of your control. Doing so simply passes the buck and does nothing to acknowledge or deal with risk.

The mature approach is to admit that inadvertent mistakes happen. No one does this on purpose and the goal is *not* to find out who goofed, but instead to find out what's weak or perhaps even incorrect in a design.

Third, you must keep in mind that you are looking for causes of *failure modes, not causes of effects.* You want to know what design factors could cause a function to be disrupted, and you don't even need to know what effects could come to pass once the mode appears in order to brainstorm possible or likely causal factors.

Finally, you need to remember that causes are—in an opposite sense to effects—events that take place *before* a failure mode arises.

If you can keep these ideas in mind and follow this guidance, you're likely to find that Step 5 is relatively straightforward and direct—and can be completed relatively quickly. However, you do need to deal with the issue of occurrence, and that can be a bit more involved than brainstorming likely causes.

DEALING WITH MULTIPLE CAUSAL FACTORS

Just as we did for effects, we must come to grips with the fact that multiple causal factors can be at work in any complex failure scenario. And, in doing this, we will need to recognize a fundamental limitation of DFMEA.

To see this in the most graphic terms, let's discuss one of the more controversial failures in aviation history: the in-air breakup of TWA Flight 800 over Long Island Sound in July, 1996.

If you want to believe in any of the alternative theories, such as an errant missile strike from the U.S. Navy, a terrorist attack with shoulder-fired missiles, or a bomb planted on board, you are certainly entitled to hold that opinion. However, the evidence gathered and published by the U.S. National Transportation Safety Board (NTSB), compiled in a 341-page publicly available report,[1] tells a very different (and more plausible) story.

Without describing all of the evidential details, NTSB concluded that the plane exploded, with the explosion occurring in the center wing fuel tank. The tank was only partially filled with fuel, and the vapors from the fuel mixed with the air in the tank, at a critical moment forming something approaching a stoichiometric mixture. As the plane climbed to altitude, the pressure and temperature in the tank and the mix of fuel vapor and air reached a condition in which the content of the tank became explosive.

This isn't a common event, but it did happen from time to time before the TWA 800 accident. It just depends on how much fuel might be in a tank at a given moment in time. However, it wasn't considered a critical issue up until then because an explosive mix won't detonate unless there's an ignition source. And the only ignition source in fuel tanks is the fuel sensor, or "fuel quantity indication system" in aero-jargon. But these sensors use very low voltage and current—not enough to trigger an explosion.[2]

After patiently and deductively eliminating more than a dozen possible causes—including those favored by conspiracy theorists—the NTSB reasoned that the most likely "root cause" was some transfer of electrical energy from outside the fuel tank to the fuel sensor system, which in turn set off an electric arc and exploded the center fuel tank.

---

1. In fact, this report is fairly critical of Boeing's FMEA and FTA assessments of the potential for fuel tank explosions.

2. The maximum power in the sensor circuit is less than 10% of the minimum needed to theoretically explode a stoichiometric mix of fuel and air—and less than 5% of the amount that the NTSB found was actually needed in lab tests performed as part of the investigation.

However, several things probably had to happen for this to be possible. First, wire bundles had to be designed, assembled, or modified through maintenance in a way that put a high-voltage cable close enough to wiring for the fuel sensor to transfer energy. Second, transferring energy from a high-voltage cable to the fuel sensor wiring required some sort of mechanism, and a short was by far the most probable cause.

This condition in turn required something to occur within the wiring bundles on the plane, particularly the insulation and separation or segregation of wires, that would allow a short to be initiated and then transfer power to the low-voltage wiring of the fuel sensor system. In the end, the evidence suggested that this could have occurred because the insulation on the wiring in this section of the plane—as well as wiring in other sections—had deteriorated over time. The Boeing 747-131 model in service on Flight 800 was one of the first produced and was placed into service in late 1971, so the wiring on the plane was at least 25 years old at the time of the incident.

Much of the wiring used on the 747-100 series aircraft—and much of the wiring used on aircraft in service today as well as on the space shuttle—is insulated with a material called Kapton.[3] This material has been shown to decay over time, and this decay is implicated in the Swissair Flight 111 flight disaster (a McDonnell Douglas MD-11) over Nova Scotia in 1998 and is suspected as a potential causal factor in the 2009 crash of Air France Flight 447 (an Airbus A330) off the coast of Brazil.

When Kapton decays sufficiently, the potential for a short or arcing increases substantially; there was a notable amount of this type of insulation deterioration found on wiring recovered from the TWA 800 wreckage.

So, what's the chain of events that led to this terrible tragedy? At a minimum, the underlying design factors—not to mention possible maintenance and inspection procedures, manufacturing errors, and other non-design factors, such as the ambient temperature and pressure in the fuel tank—had to include these causal factors:

1. The center fuel tank had to be designed so that fuel vapor and ambient air could mix freely and not be suppressed.

2. Fueling procedures had to be specified that caused vapor and air to be mixed in a way that was explosive.

---

3. Boeing stopped using Kapton in 2002 in part due to the problems discovered in both the TWA and Swissair accidents. In fact, one of the reasons that NASA is retiring the shuttle fleet is that each shuttle has 140 or more miles of Kapton insulated wire. But a majority of the commercial jetliners in the air today still have Kapton insulated wiring in much of their electrical systems.

3. Wiring bundles had to be designed so that energy from high-voltage wiring could be transferred to the fuel sensor wiring.

4. Wiring insulation that would decay over time had to be specified—and that decay had to result in a short that would transfer power to the low-voltage fuel sensor system.

Considering design factors alone, the probability of getting all of these factors aligned in the wrong way is unbelievably low, yet all 230 people on board died because that seems to be what happened. The chain of events that led to this accident is so long and convoluted that it can seem impossible and makes the alternate conspiracy theories that still roam the Internet seem more persuasive than the evidence suggests they really ought to be.

Some of the probability estimates for individual events in this chain are at a level of 1 in a billion or so—and several events of this type all had to line up, yielding an extraordinary improbability.

Would FMEA have discovered this chain of events? The answer is simple—it would not. In fact, the NTSB found that Boeing's after-the-fact FTA study of this chain wasn't particularly sound, either.

However, a bigger question exists: how many of the four cause–mode–effect chains that are design related alone would have had to be broken to have prevented this accident? The answer is simple: one. If any of these causal factors had been made even less likely or, perhaps, nearly impossible, the TWA 800 accident might well have been averted.

Given this example, we are forced to conclude that DFMEA will not identify a complex chain of cause-and-effect events of this type. However, a series of DFMEA studies, with proper scope and depth of analysis, would likely identify one or more of the causes in this chain. And, if just one of those causal factors can be made extremely unlikely, the entire chain of events will become a once-in-the-life-of-the-universe probability.

Risk will remain, though, and the acceptance of this risk needs to be understood. But the need to trace the entire potential chain of causes and effects is not only difficult to do, but would take a huge and impractical effort to achieve in any design process.

We'll conclude the discussion of "how many causes should be considered" after taking a bit of a dive into the issue of occurrence. However, at this point, we'll simply suggest that you can't really have an effective DFMEA if you try and list every possible causal factor for each failure mode.[4]

---

4. Again, this is a place where the AIAG reference manual on FMEA isn't as sound as it might be—it strongly suggests that all possible causes be listed for every mode. We'll see why this is impractical later in this chapter.

UNDERSTANDING OCCURRENCE

The most common difficulty in Step 5 is understanding and properly applying occurrence ratings. Occurrence—which only applies to causal factors and does not directly apply to effects—is not a clear-cut concept in DFMEA. It's much easier to understand in PFMEA, but in DFMEA occurrence becomes something of a conditional probability.

To grasp this, we need to start with an operational definition for occurrence:

> Occurrence is a numerical rating of the probability that a given cause will arise and will result in a specific failure mode.

> Occurrence is rated from 1 to 10 using standard tables.

What this really means is that occurrence is a probability estimate that asks how often a causal factor will be translated into an error that is seen in the final product. The challenge in rating occurrence is that a design team almost never has enough information to decide on the underlying probabilities in any kind of analytical way.

Most occurrence tables provide an outcome in terms of number of errors per opportunity. For example, in the AIAG table for DFMEA, a rating of 10—the worst possible rating—is any incidence of failure greater than 1 per every 10 vehicles. A rating of 2 is any incidence of failure less than one in a million (better than six sigma) and a rating of one is reserved for impossibility. (That's a bit of a crazy standard; it's really, really difficult to conceive of a cause–mode–effect chain where a rational causal factor can be made impossible through preventive control because there's always a weak element somewhere in the causal factor being considered. But that's the way the AIAG describes an occurrence rating of 1.)

Other tables, like those provided in the appendix of this book, have different rating tables for different scales of operation. In low-volume manufacturing, for example, even conceiving of a one-in-a-million probability can be difficult or even impossible. A one-in-100 probability may be the very best that can be realistically anticipated.

Nonetheless, you need to make an estimate of occurrence. And this can be quite difficult unless you have a well-validated reliability model. What does this really mean?

To begin, we need to understand that occurrence is not an estimate of how often a design error is made. It's not about how often a tolerance is calculated incorrectly; instead, it's an estimate of how often an incorrect calculation results in a disruption of function in an actual product. That's much harder to estimate in most cases.

Next, we need to acknowledge a simple yet often misunderstood fact of life. It's almost unheard of to design something so that it fails every time. This could be done deliberately as a demonstration of a principle or to prove a point, but it almost never happens, even in very low-volume businesses. It's much more common to design something that, when made according to specifications, goes wrong occasionally or infrequently.

In other words, the probability of a causal factor leading to a failure mode is never 100%. If you specify a material with insufficient strength to support a load, you are not likely to get this so wrong that every piece produced would fail. Instead, it would be much more likely that only a few pieces, just those where the material strength happens to be at the low end of the specified tolerance, would fail.

The cause—incorrect material strength—would always be present and would be a design decision. However, the number of times that this specification would lead to failure would be relatively small, meriting a relatively low occurrence rating.

For most functions, there are one or perhaps two design factors that are important or even critical for the design to be successful. In nearly every case, these factors must be specified with limits, an acknowledgement of the fact that manufacturing is never perfect. A tensile strength for steel, for example, is either called out as a minimum or as a range with a minimum and maximum acceptable value.

If you asked a steel mill for a precise and exact strength on every piece produced, they'd either laugh at you behind your back or turn down your order. So, you have to expect—and design in—allowances for variation in manufacturing.

It's important to emphasize that this does not violate our assumption that manufacturing will be carried out flawlessly, because we continue to assume that all steel will be delivered *within the specified limits,* whatever those limits might be.

In sum, critical material properties (or other design characteristics) that affect a function will be distributed over some range. There will then be some finite probability that this primary property—either alone or in combination with other distributed design characteristics—will yield a result that does not meet the overall performance requirement for that function.

In most cases, these properties could well be distributed normally, falling within the distribution of a bell-shaped curve that is familiar to most engineers.[5] For a normal, or Gaussian distribution, we can predict what

---

5. Even if the distribution isn't normal, the principles explained in this discussion still apply in most circumstances. Most non-Gaussian distributions still

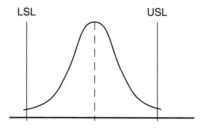

Figure 7.2   Normal distribution with specification limits.

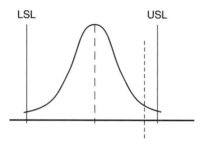

Figure 7.3   Typical distribution of actual manufacturing results with "trouble point" noted.

percentage of outcomes will occur by looking at the area under a segment of the curve—the boundary lines are the *lower specification limit* and *upper specification limit* (see Figure 7.2).

Using the simple steel example started above, if we specify steel that is grossly weak for the design we are considering, we might find that 95% of all parts made with this steel would have insufficient strength, and the part might fail. In this sense, a critical value would exist at the value indicated by the dotted line shown in Figure 7.3—only 5% of the area under the curve would be above this line, so only 5% of the delivered steel would be likely to have the strength needed to survive.

exhibit the same approximate behavior as Gaussian distributions, even if the underlying math isn't equivalent. There are a few that are quite odd (like bimodal distributions) for which the specifics are different, but those are unusual and quite rare for most design activities.

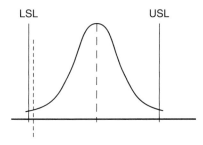

Figure 7.4   Distributed results in manufacturing with critical value at 0.5% of specification at trouble point.

However, that's not a common outcome—an opposite outcome would be that 99.5% of all parts made from the specified steel might have enough strength to survive and only 0.5% might fail.[6]

In that case, the "critical value" point might shift far to the left, so that only 0.5% of the area under the curve would be between the lower specification limit and the "trouble point" where the strength would be insufficient (see Figure 7.4).

If you knew these factors—and had data to back up these conceptual pictures, you'd be able to rate occurrence quickly and accurately. In reality, you rarely know these values, even in the simple case just described, without detailed simulation and/or calculation of some type.

The truth is that you hardly ever know this kind of information; it's sometimes not clear which product characteristics will have an impact on a function, and determining the underlying reliability considerations that could make this information apparent can be too difficult or time-consuming for all but a small number of functions in a given project.

What, then, are the alternatives for estimating occurrence? The first and easiest is to simply make an educated guess. This isn't complicated, but it doesn't really give you a sound basis for evaluating risk and planning controls—the main reasons that DFMEA is valuable and important.

One of the better techniques for shifting from a wild guess to a more structured guess is a table developed by AIAG—even though the basic idea is presented in the *Production Part Approval Process Reference Manual* but isn't contained in the latest *FMEA Reference Manual*.

---

6. You'd have a hard time finding this with physical testing—more about that in Chapter 8.

**Frequency of failure based
on historical evidence**

Type of evidence	Low	Medium	High
Actual experience	1	4	7
Surrogate experience	2	5	8
Assumption	3	6	9
No background			10

Figure 7.5   Modified AIAG table for estimating occurrence.

Over the past few years, I've modified the table AIAG developed and made it available to teams as a basis for estimating occurrence (see Figure 7.5). It's not perfect, but it's far better than just guessing a number based on notional considerations. This table does not replace a standard occurrence table with numerical occurrence limits, but can be used to estimate occurrence when reliability models are not available.[7]

This table relies heavily on organizational experience as a stand-in for fully developed reliability data. As such, it isn't as good as a serious reliability analysis, but I've found that it provides a repeatable and reproducible basis for assessing risk. And, after all, that's what DFMEA is all about.

In practice, I've also found that occurrence ratings higher than 5 aren't very common in most product design projects.[8] To start, it's rare that an important design characteristic, function, or feature is included if a design team doesn't have some experience with the conceptual underpinnings of the issue. It's even more unusual to choose a design feature or characteristic that has suffered a high rate of historical failure.

Looking at the table in Figure 7.5, that would make any rating higher than 6 impossible unless the design was radical and had no relationship to previous designs or even conceptual assumptions based on some kind of analytical model. Overall, you really have to be "fishing in the dark" to choose a design feature or characteristic that would have a high historical failure rate or for which you had no background at all.

When all is said and done, though, you need to come up with a number. Choose an appropriate table—again, there may be such a table for your industry, or you may wish to use one of the tables provided in the appendix.

---

7. Of course, the rating of 1 in this table is inconsistent with the occurrence tables in the AIAG FMEA reference manual—an artifact of how AIAG reference manuals are created, by teams of loosely connected volunteers.

8. This isn't necessarily true in PFMEA, however.

And, if you find you don't know enough to guess an occurrence using a standard table, then you can always fall back on the supplemental table shown in Figure 7.5.

SELF-DECEPTION ABOUT OCCURRENCE

Engineers have an amazing ability to fool themselves when it comes to estimating occurrence, particularly when using historical data. Once again, an example from the aerospace world will help illustrate how powerful this tendency can be.

This time, we need to consider the *Columbia* disaster that claimed the lives of seven NASA astronauts in early 2003. In this accident, debris from the large orange external fuel tank, probably consisting of insulating foam and ice, broke loose during the ascent of the orbiter on launch. This debris then poked a hole in the thermal insulating material on the leading edge of *Columbia*'s left wing.

Upon reentry into the atmosphere, the hot plasma that forms on the leading edge from friction due to wing drag in the upper atmosphere then flowed into the internal wing area, melting the underlying metal structure and causing the orbiter to disintegrate. The fate of the seven crew members was sealed, and the debris was subsequently scattered over several states.

The possible causal factor, of ice or foam coming off the fuel tank and striking the leading edge, was something that NASA had known about from the start of the shuttle program. And, since the ill-fated flight of *Columbia* was the 113th shuttle mission, you would think this issue was well understood. In fact, three earlier missions had suffered damage from debris from the fuel tank, but that damage didn't result in sufficient harm to cause any real problems—yet another illustration that you can not easily predict which effect might arise once a function (that is, insure thermal protection of the leading edge of the wing) is disrupted.

Further, a simulation program had been developed by NASA to gauge this risk and found that there was a 1 in 15 chance of serious impact of some sort—but not necessarily an impact that would disable the orbiter. However, after many flights, the actual damage found suggested that the occurrence of a serious but nonfatal puncture would be less than this—it was about 1 in 30 or so.

If the additional and conditional probability of a nonnegligible puncture of the leading edge of the wing that would be potentially fatal was factored into this, the probability would probably reach a 1 in 200 value. Of course, there was no verified reliability analysis to this effect, but this is a value that is believable.

This *Columbia* mission, designated STS-107, was the shuttle system's 113th launch and *Columbia*'s twenty-eighth launch. So, given that all of these prior launches occurred without serious damage from tank debris, what is the probability that significant debris damage would occur on this particular flight?

Because there's no cumulative impact of causal or usage factors, the probability on this flight would be approximately the *same as it was on any flight,* or about 1 in 200 or so.[9]

It's not more probable because it didn't happen before, although some people do think that this means "your number is up" because you haven't seen the fatal effects in so many previous flights. Further, it doesn't mean that it's less probable because you've done this "experiment" 112 times and never had it go wrong, although that kind of thinking is extremely common in many design organizations.

But—what did program managers and engineers believe? In fact, there are at least three daily report comments, written by NASA engineers during *Columbia*'s mission, that acknowledge the debris contact during launch and discuss and then dismiss the potential for serious damage. Many people have questioned why the highly competent NASA staff reached these conclusions.

As Henry McDonald, former director of NASA's Ames Research Center, told Congress in testimony after the disaster, well-intentioned people forget that "if you have a 1 in 100 chance of risk of an event occurring, the event can occur on the first or last opportunity and there's an equal probability each time."

He further went on to state that NASA's perception seemed to be "that if I've flown 20 times, the risk is less than if I've flown just once."

In fact, unless there was a significant change in design and testing of the shuttle system, the risk remained the same each and every flight. If the odds are 1 in 100 and you've had 99 successful flights, that doesn't mean you are destined to fail on the 100th flight. And, if you have two failures in

---

9. This isn't completely accurate; small but important changes in the fuel tank insulation material, as well as launch procedures, had been made over the years. In addition, *Columbia* was the oldest of all the orbiters and so was heavier and had somewhat less sophisticated technology than later orbiters. So, the 1 in 200 number is even less valid—but the general principle is still the same.

the first ten flights, that doesn't mean you are less likely to have failures on future flights because you've experienced failures already.

I can hear most engineers now saying that they know all of this and they fully understand that the odds of something happening in a given case aren't changed by what happened in earlier, independent cases.

And I know just as well that the overwhelming majority of engineers don't behave in a way that is consistent with this intellectual understanding. In fact, it's all too common that engineers fall victim to believing what they want to believe. Many people believe that engineers, as disciplined thinkers, are less likely to be victimized by these kinds of fallacies, but it's really amazing how often this occurs. I've done it myself, and I know how easy it is to think that prior positive experiences actually change the underlying probabilities of events.

As an illustration, let's consider a hypothetical project. As part of this project, a comprehensive test protocol has been developed, and this protocol requires that twenty separate samples of some piece of hardware must be subject to an involved durability test. The first eighteen samples meet all the criteria of the test without difficulty. Then, during the nineteenth test, a failure occurs late in the test.

Will the design be reexamined and then altered to address what went wrong during this test? Will the previous eighteen successful tests then be discarded because they no longer represent the latest level of the design?

It might be true that the design will be altered and the testing protocol repeated from the beginning. In a perfect world, that would certainly be a probable outcome in a situation like this, and fewer after-release design errors would enter the world.

On the other hand, I'm willing to bet that at least forty-nine out of fifty times, the failure in the nineteenth test will be disregarded, explained away by some odd or known problem with the test equipment, or some obvious (or not-so-obvious) out-of-specification characteristic found in the tested prototype. Something, anything, will be found to justify treating this failed test as a statistical outlier, or "flier," and thereby allow the project to move forward without the huge time and cost needed to redesign and retest the design.

Of course, it may well be that the test is somehow tainted. In my long experience, though, I've found that it's much more likely that something else will ensue.

The evaluation of the results of the nineteenth test will likely be affected by the perception that 18 previous tests must be worth something. For most people, the emotional desire to see a successful outcome for the project will also weigh heavily in the interpretation of test number 19, and that will also—in most cases—tilt the scale toward disregarding or minimizing the significance of the results from that test.

That's really not much different than NASA's perception of the risk of tank foam hitting the orbiter during launch. And I need to emphasize that this is not intended as criticism of NASA's highly dedicated and technically skilled people. They care deeply about what they do and fell into this "common sense" trap because they *are* smart, dedicated, and capable.

This is a very sophisticated and widespread type of self-deception that can easily appear in any team's attempt to interpret previous testing. You need to understand the underlying reliability issues that affect occurrence, which is something that makes the use of almost any occurrence table in DFMEA a fragile proposition.

Still, in the end, you need to make an estimate about occurrence. Do the best you can and keep in mind the cautions I've tried to raise while you do so.

TECHNIQUES FOR CAUSE AND OCCURRENCE ANALYSIS

To actually work through the next two columns on the main worksheet, you can employ the same basic modified brainstorming strategy that I suggested for effects and severity analysis.

First, we'll be skipping the very next column, the one titled "classification." We'll return to this in Chapter 9 and we'll then understand why this column has been temporarily passed over. For now, just ignore it.

Next, look at the existing worksheet and first *hide the columns for effects and severity.* Why should you do this? Because you are looking for *causes of modes,* not causes of effects. If you are doing this on a spreadsheet, you can temporarily "hide" these two columns. If you are using paper, an easel, or a whiteboard, create a paper or cardboard mask to block your view of these columns. While you're at it, you can hide the "classification" column as well.

Never forget, DFMEA can't accurately link specific causes and effects. It can only link causes to a specific mode and effects to a specific mode. The specific interaction of causes and effects is beyond the capacity of FMEA techniques.

With three of the columns hidden, your spreadsheet now looks like Figure 7.6.

Now, brainstorm likely potential causes for each mode, making sure that you consider only design errors and not errors outside of the design

Item/function	Functional requirement(s)	Potential failure mode	Potential cause(s)/ mechanism(s) of failure	Occurrance
Locate rod to bushing ring	14.5 mm from bushing face to rod OD	Rod partially located		
		Rod loses location over time		
		Rod incorrectly located		

Figure 7.6   The DFMEA worksheet with effects and severity hidden—ready for brainstorming of causes.

process. If you can think of more than one cause, look at the list and select the top cause in terms of occurrence—but don't try and rate the occurrence just yet. Then, look at the rest of your list. If any of the other causes are close to the most likely cause in terms of occurrence, you can add these as well.

This will cause further splitting of each row on the worksheet, and you will have to keep this in mind when you are assessing risk in Chapter 9. (It also makes automating a spreadsheet with macros a bit more complex.)

However, I once again urge caution. Listing all of the causal factors for each mode can be a monumental task, and, if some of the causes have vanishingly low occurrences, you really need to ask what benefit you will get by adding these to your analysis.

Most of the actionable outcomes from a sound DFMEA study will come in the form of design changes that will either eliminate potential causes or reduce their occurrence. At the end of the day, how many changes—or corrective actions—will you be able to undertake and implement?

In the case of the trailing arm weldment—a simple three-piece component—we found at least seventeen functions. If we figure that each function has two to three modes on average, this means a total of forty or forty-five modes are reasonable. It may be more, but there are unlikely to be fewer. If you found three causes for each mode and you only considered one effect for each mode, this would mean perhaps 150 cause–mode–effect chains commanding your attention.

Realistically, will you address all of these? In any but the most intensive or risk-averse projects, I think the answer is clearly no, particularly when you realize that the weldment analysis is just one of more than a dozen subsystem analyses in the rear suspension system. There are only so many hours in a day and so many days in a project and you must make intelligent decisions—as opposed to elegant and rigorous decisions—as an engineer.

If you are working on new technology, you could include many or even most of the causes you can think of. Do remember, though, that each one of these will need to be assessed for occurrence—and then for design control issues and overall risk. That adds considerably to the overall effort required to complete the DFMEA process.

When you enter every possible cause, it becomes likely that one or more of the few truly significant and riskiest C-M-E chains will be drowned out by the trivial many. In any system of communication—and DFMEA is a form of communication—too much information can be a hindrance, as critical issues are overwhelmed by mundane and insignificant details.

This requires a value judgment, of course, which is a bit of a departure from the "deductive" approach this book advocates, but it is, I believe, a necessary one. Moreover, if you use occurrence as a "sorting" tool for considering only a couple of causes, you are still engaged in a highly deductive methodology.

If you develop a list of five or even ten causes, you can make this value judgment more deductive and less inductive by considering the relative occurrence (without using any tables) for the entire list. Just rank-order these causes and see if there's a break point between more-likely causes and less-likely causes. If you do that, you will, in nearly every case, limit the number of causal factors that are worth worrying about to two or at most three.

With that in mind, after you've listed all of the causes you think are worth considering, then and only then should you attempt to rate occurrence for each cause. Most of the concerns that impact the use of severity tables are again relevant, including the fact that no table is perfect; when in doubt use the higher number, stick with just one table—or perhaps two if you use the supplement table in this chapter as well—and recognize that your pace should be quick. If you are spending too much time debating occurrence, you are probably not using the table properly.

And, as recommended for severity, if there's a debate about occurrence, choose the higher value unless there's a great disparity between viewpoints. If there are significant differences of opinion—more than one number on the occurrence table—then a much more detailed discussion is probably in order. This is particularly true in DFMEA, where the probability of an occurrence greater than 5 is extremely low.

All in all, cause-and-occurrence analysis should be brisk and lacking in drama. After you've gotten a grip on rating occurrence, it should be one of the simpler steps in the entire DFMEA process. If you spend more than a few dozen seconds discussing each rating, you are probably falling into the trap of "analysis paralysis" where the debate itself takes on more importance than the resulting assessment of occurrence.

SUMMING UP: STEP 5

Although assessing occurrence can be difficult, it should not be complicated to brainstorm a solid and actionable list of causes. As long as you remember that the causes have to be design-driven and must occur before a function is disrupted, this is usually a clear-cut exercise.

As we tried to do in Step 4, the underlying theme is to keep things on a simple and direct basis. One cause for each mode is necessary, two may be worthwhile, but, unless you have a large number of high-occurrence causes, developing, listing, and tracking more than two causes for each mode yields diminishing returns.

And, of course, keep working on a column-by-column basis.

So, to approach this step you simply need to brainstorm causes for each mode. Blocking out the effects and severity columns from your previous work will help ensure that you think only about things that arise before a function is disrupted. After listing a few causes for each mode, enter those causes that you think are most likely.

Once causal factors for each mode have been entered on the worksheet, you are ready to tackle occurrence ratings. Pick the right table for your project or business and, to the best of your ability, try and use this table. If you have difficulty, you can always fall back on the supplemental table in this chapter for assistance.

Again, this can all be recapped with a simple diagram (Figure 7.7)—almost, but not quite identical to the diagram used to summarize Step 4.

By following this diagram for the weldment project, our worksheet is filling out. Keep in mind that cause and occurrence ratings in this example are hypothetical; these results would vary from organization to organization depending on the design group's experience, previous history, and any of the other factors that drive causes and occurrence ratings.

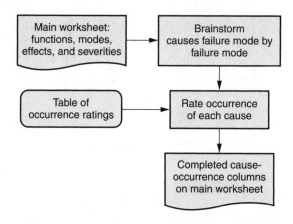

Figure 7.7   Summary of Step 5 activities.

If, as a reader, you are unhappy with any of the supposed entries in this example worksheet, by all means create your own. As an author, I'm certain that many experienced chassis engineers in the automotive industry would generate a somewhat different result—but I'm equally certain that the results presented here are at least plausible and understandable for most readers.

If you've hidden the effects and severity columns when brainstorming causes and rating occurrence, you should allow these columns to be seen before moving on to the next step.

Figure 7.8 shows the worksheet for the weldment example, updated with causes and occurrence ratings. In our hypothetical example, I've rated the occurrence of a bad welding callout very high, since this kind of part has historically been made (at least by our imaginary company) by forging. That's a high occurrence, but because there would likely be doubt about this, I chose a higher number.

Item/ function	Functional requirement(s)	Potential failure mode	Potential effect(s) of failure	Severity	Classification	Potential cause(s)/ mechanism(s) of failure	Occurrence
Locate rod to bushing ring	14.5 mm from bushing face to rod OD	Rod partially located	Rod–ring interface fractures; loss of vehicle control	10		Angular relationship between rod and ring incomplete	2
		Rod loses location over time	Steering precision is diminished	7		Rod section modulus insufficient	3
			Rear wheel traction is decreased	7		Incorrect weldment callout	5
		Rod incorrectly located	Premature tire wear	5		Incorrect suspension geometry provided by system engineering	2
						Angular relationship between rod and ring incorrect	3

Figure 7.8 The DFMEA worksheet with functions, modes, effects, and causes.

8

Step 6—Controls and Detection

At long last, we're ready to develop some discerning information from the DFMEA process. It's very likely that you've learned a great deal about the design of your project up to this point, particularly regarding function and the specific requirements for each function.

However, we are now ready for some very important questions:

- What will we actually do to verify the design?

- How effective do we think these actions will be?

We have all of the information that we need to do this. We just need to get on with the effort. In doing this, however, we need to make sure that we do everything we can to minimize the need to rely on physical testing and to avoid falling into the trap of a repeated trial-and-error method of verification. Instead, our goal is to develop a design verification plan with strong emphasis on prevention as opposed to detection of design flaws (see Figure 8.1).

In many ways, our goal is to emulate the Wright brothers and aim for a more analytical approach to verification—an approach that will be less costly, more revealing, and offers the potential for better control of project timing.

As usual, we need to start with an operational definition, this time for *control*. For DFMEA:

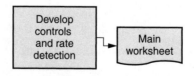

Figure 8.1 Step 6 of DFMEA process.

139

A design control is an activity that reveals or makes visible a particular failure effect or failure cause.

It is rare that a failure mode can be observed or otherwise sensed, since a mode occurs at an instant in time and is not a sensory event. Before a mode occurs, a cause might become perceptible, while an effect can only be sensed after a function has been disrupted.

Ultimately, all of the controls that may be developed in this step should be included in a formal design verification plan. By doing this, you will be applying the lessons that DFMEA can teach in a way that promotes a constructive and compelling application of the chain of verification.

Because there are two different categories of controls, we need to examine the general activities that take place in the design process itself. In any design project, there are certain inputs. These can include marketing requirements, cost targets, functional specifications, regulatory requirements, and general as well as specific customer preferences.

These inputs are then transformed by dozens or even thousands of design decisions to create design output. Design output includes final drawings, specifications, and all of the information that a manufacturing organization will need to plan, develop, and launch a manufacturing process that will fabricate and assemble the designed product.

Once again, a simple diagram can be constructed to show this process (see Figure 8.2).

The first type of control we need to consider is called a prevention control. *Prevention controls* are aimed at design process inputs and causal factors—the left-hand side of this diagram. In almost all cases, prevention controls are based on analytical assessments or comparative examination.

Prevention controls include simulation, computer-aided engineering studies, calculations, and related activities. Prevention controls can also include comparison against historical results, formal design reviews, or comparison to industry standards, supplier specifications, or organizational standards.

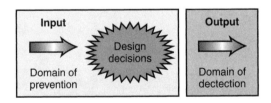

Figure 8.2 The relationship between prevention and detection controls and the design process.

In almost all cases, design-oriented prevention controls are intellectual activities rather than physical activities. You use your brain (as Orville and Wilbur Wright did) to examine the most important issues for each function to visualize or simulate what might go wrong when the concept design you've been considering is expanded into a fully-developed and detailed set of design outputs.

In contrast, *detection controls* are aimed at process outputs. Since the only way you can really see if a failure effect arises is to build and test prototypes, this means that detection controls consist of physical tests.

Sometimes, this division of control activities can appear confusing or even incorrect. If we think about the Wright brothers, they conducted a fair amount of physical testing using a wind tunnel. At first glance, this seems to be a type of control activity because they were attempting to determine critical design inputs, namely values of lift and drag for different airfoils.

In reality, though, the Wrights' wind tunnel experiments were not about verifying design issues. They didn't place scale models of their gliders in the wind tunnel; instead, these experiments were classic research and development activities, intended to discover fundamental values of physical constants. Because these experiments never included actual design examples, they really weren't design controls at all.[1]

A prevention control, when applied to inputs, is used to question the soundness, accuracy, or applicability of the input, not to develop the input *per se*. Creating or developing an input is another type of activity altogether and could include market research, physical research, or any type of information gathering.

If this is too confusing or doesn't ring well for you, then don't hesitate to think of some physical testing as some sort of prevention control. As long as the effort is aimed at inputs and causes—and not aimed at effects— the activity will be a prevention control. If the activity is aimed at seeing effects, then the activity will be a detection control.

WHY PREVENTION CONTROLS ARE IMPORTANT

Too many engineering organizations seem to believe that design verification means "test plan." Certainly, testing—as embodied in detection

---

1. This is another example of why research is difficult to do in the midst of a commercial project, as explained in Chapter 2. It's far better to do research as a separate precursor to directed design work.

controls—is a major and important element in design verification. Nevertheless, we really need to understand why prevention controls are so important and why detection controls are limited.

Although we discussed this in some detail in Chapter 2 and we have added ideas throughout this book, we need to reemphasize this and look at a few additional reasons why prevention controls are so essential for competitive product development project management.

The first and most important thing we need to recognize is that physical testing has restricted value. In most cases, so few samples are tested that the statistical significance is terribly, even laughably, low. To explore this, let's once again return to the auto industry and consider the complex case of the Ford Explorer–Firestone Wilderness AT tire debacle that surfaced at the turn of the twenty-first century.

The entire story is complex, and there are many aspects that go far beyond what we need to know for our purposes. In sum, the first generation Ford Explorer, a body-on-frame[2] sport utility vehicle (SUV) with a relatively high center of gravity, was, in most cases, outfitted with Firestone Wilderness AT tires.

In a small but very meaningful number of cases, the tread of the Wilderness AT tires would separate from the tire carcass. And then, in even fewer cases, this would cause a loss of control for the driver. In some of these cases, the vehicle rolled over—and, in fewer cases yet, occupants in the vehicle were either killed or seriously injured.

Various estimates suggest that more than two hundred people were killed and more than a thousand were seriously injured from this chain of events. In many ways, the web of cause-and-effect relationships in this tragic situation is even more complex than in the TWA 800 disaster. Indeed, to this day, Ford and Firestone—two companies linked by history and family heritage—have not agreed on the root causes of this industrial fiasco.

No matter—let's look at just one aspect of this: testing. The number of causal factors that are involved in a rollover are substantial. To name a few, these can include the design and manufacture of the tire, the design of the vehicle (center of gravity, suspension system, weight distribution, and so on), air pressure in the tires, road conditions, wear of the tire tread, and the temperature of the road surface.[3]

---

2. This makes the vehicle more truck-like and less car-like—and generally diminishes overall dynamic stability when compared with a unibody design of the type used for nearly all common passenger cars.

3. Of course, if only one cause–mode–effect chain in this scenario can be broken, the entire chain may not occur.

The only way to assess all of these factors with a physical evaluation is to conduct a tire durability test. In a tire durability test, a test vehicle is driven at various speeds, either on a closed course or on a set public road route. The goal is to drive the vehicle until the tires wear out, which may require as much as 50,000 miles of driving. This can take eight weeks for a single test, assuming a rotating crew of drivers who are accumulating mileage 24 hours per day.

In the Explorer program, Ford (according to public testimony) conducted one tire durability test. What could be learned from one test? Let's assume that there's one chance in two thousand that the Explorer–Wilderness AT system would suffer a tread-to-carcass tire failure *and* that event would also lead to a loss of control for the driver.[4] Under these simplified assumptions, that's 99.95% reliability—and this doesn't even include the additional conditional probabilities involved with rollover and subsequent injury.

What's the chance that the failure scenario would unfold in one test? It's one in two thousand, or 0.05%, the complement of the reliability. It might happen, but finding a significant design flaw, or causal factor, with this test would be a stroke of luck of the highest order.

Now, let's turn it around and ask a different question. How many tests would be needed if we wanted to have a 90% probability of seeing the tire failure–control loss scenario just once? Since this is an either/or proposition, we can model this using a binomial distribution. If we wanted to have a 90% chance of seeing one failure, given a one in two thousand underlying reliability, almost *eight thousand* individual tests would be required.

That's never going to occur in a vehicle development program.

Still, 99.95% reliability sounds quite good—until you realize that Ford sold a bit more than 2 million Explorers of this particular design. That equates to about one thousand Explorers that may actually roll over—and, at least in general magnitude, that's approximately what happened.

As bad as this is, the vagaries of testing make this simple-minded analysis even less valid. *Car and Driver* magazine[5] tried to force an Explorer to roll over. They fitted a well-used Explorer vehicle with a competition roll cage and had race-style seat belts installed to protect the driver. Then, at various speeds, up to and including seventy miles per hour, they flattened one of the rear tires using a rapid-action deflation valve.[6]

---

4. That's at least a plausible number—perhaps a bit too high, but close to the general order of magnitude—according to unpublished sources.

5. *Car and Driver* (January 2001).

6. This valve would deflate the tire in less than 1/3 of a second.

They could *not* cause the SUV to tilt enough to start a rollover reaction. In fact, the test driver was able to bring the vehicle to a safe stop from seventy miles per hour when applying the brakes as hard as he could—and without touching the steering wheel. The deflation of the left rear tire (the tire thought to be most likely to cause a loss of control) was insufficient to induce any real problem.

Of course, there are a lot of things that happened in this test that don't fully simulate real-world usage. First, the driver was a skilled professional vehicle evaluator and amateur race driver who didn't panic. That means he didn't yank the steering wheel to one side, causing the vehicle to veer off the road. Second, the road surface was smooth and well maintained (they used a drag strip raceway to conduct the test) and it didn't have bumps, potholes, or cracks. Third, the actual tires on the vehicle were Goodyear tires, not Firestones (these were the tires that were installed on the vehicle when the vehicle was purchased in a private transaction). Fourth, the tires were inflated to the proper pressure and had not been repaired for a puncture, factors that Ford maintained were part of the problem. And finally, the test was conducted in the late autumn near Detroit—rather than in a hot desert clime where road surface temperatures can exceed 140° F.

So, what did Ford learn from the one tire durability test they conducted? To be concise and accurate, they learned that one prototype vehicle, fitted with what were likely prototype parts, did not suffer any major tire–suspension system interaction failures while driven over a prescribed test course. That is a long, long, *long* way from suggesting that the tire–suspension system was verified in any meaningful way.

What's the alternative? The U.S. National Highway Traffic Safety Administration (NHTSA) has developed a relatively simple calculation that predicts the propensity for rollover, based on general vehicle dynamics. They've augmented this with tests of production vehicles, and most of the major vehicle manufacturers have developed even more sophisticated simulations that predict rollover characteristics as well.

In short, there are both prevention controls—simulation—and detection controls—tests, such as the "J turn" test, a maneuver that is more of a movie stunt than real-world replication, that NHTSA conducts—that are used to evaluate rollover tendencies. None of these controls are perfect, but it is clear that the incidence of rollover (or occurrence, if you prefer) has been gradually and steadily decreasing as the sophistication of the simulations used has improved.

If we step back and think about what all of this means, we can see that testing alone can't tell the whole story. In fact, most of the test results that have been obtained have been used to add details to the rollover simulation programs rather than to verify designs *per se*.

This is no different than what happened in the Apollo I tragedy. The capsule was tested, tested, and tested some more. It was even subjected to a "design certification review" and passed (well, at least it passed on the second attempt). The capsule still suffered a catastrophic failure sequence.

Let's sum up the major reasons that testing is, under the best assumptions, a less-than-optimum way to verify designs:

- Most test programs have low statistical significance. Confidence levels exceeding 70% are rare in most projects.

- Tests rarely replicate real-world conditions. In the case of vehicle rollover, this is particularly true. Few drivers ever execute a "J turn" in the life of a vehicle. Testing with all of the possible environmental factors—snow, rain, worn and/or underinflated tires, dusty roads, hot and cold pavement, broken pavement, driver error—isn't part of any comprehensive, formal test program for automobiles. While each of these conditions is part of vehicle durability testing, the confluence of factors that may be needed to see design flaws related to rollover isn't combined into one evaluation for rollover propensity.

- As noted previously, testing simply can not replicate the limit-of-tolerance conditions for any but the most simplistic situations. As we saw in discussing occurrence, it's probably not even possible to create test samples at the limits of material properties in most projects, and actually assembling test units that would have a long, concatenated string of product attributes with minimum values for important parameters is simply too complex to even consider. And yet, given the quantity of vehicles that are produced, the potential for several key factors to align at worst-case condition in a single vehicle is far greater than one in a million.

- Finally, in the auto industry, it's indisputably true: virtually all recalled vehicles were tested in hundreds of ways—and the test results were judged acceptable, at a minimum. This is generally true in almost any other industry, although no industry has the degree of public history and well-studied results that the auto industry has.

The second and most important reason that prevention is important is even simpler. Failure of physical testing in a development program is expensive and time-consuming. We looked at this in great detail in Chapter 2, and the reasons for carrying out the chain of verification in the exact order shown should be clearer than ever at this point.

However, testing isn't obsolete and probably won't be at any time in the conceivable future. Still, prevention controls are often more effective than detection controls and should be a focal point in any verification effort. This is true even when prevention controls are less effective than detection controls.

In best-practice projects, prevention controls are used extensively to effectively "debug" designs. If most of the problems with a design can be found analytically, testing and detection controls perform a simple role, namely, to confirm that previous analysis is correct. This not only saves time, it also saves considerable money when done properly.

So, the best thing to do is to create both prevention controls *and* detection controls for most or even all of the cause–mode–effect chains that might arise for a given design.

RATING DETECTION

Compared with severity and occurrence ratings, detection ratings are relatively straightforward. Some of the latest tables that have been developed for detection (particularly the AIAG detection tables) have become more complex, but in general, most detection tables range from a "certain to detect" estimate to "unlikely to detect."

In a formal sense, we can define detection as follows:

> Detection is a numerical rating of the probability that a given control (or set of controls) will discover a specific failure cause or failure effect.
>
> It is important to make sure that any given detection rating is associated with a control that is tied to a specific cause or effect.

One of the odd things about detection is the underlying language of detection. Is a "high" detection rating a big number—or is it a "high" probability of detecting a cause or effect? This can get confusing when people are discussing detection. My advice is to discuss big numbers (bad) or low numbers (good) rather than using terms like "high" and "low" when talking about detection.

One of the latest "twists" that has been introduced regarding detection is the timing of a specific detection event with respect to the overall product development process. In the AIAG fourth edition reference manual, controls that are carried out relatively early in the product development process (often prevention controls) are assigned a smaller number, while controls carried out later in the process (usually detection controls) are assigned a higher number.

In the end, detection values should be estimated for each control identified for all cause–mode–effect chains. However, because there may well be two controls for each C-M-E chain, the overall risk associated with a cause–mode–effect chain will be based on the most effective control, as long as all controls are completed as part of the design verification process.

Ultimately, if you complete both activities for a C-M-E chain, the overall risk will be based on the most useful or effective control. Even in the case where the detection rating for a detection control results in a smaller number than a prevention control, best practice means including a prevention control of some type. If a test failure can be averted by analysis, cost and time will be saved, and that nearly always makes the completion of a prevention control worth the cost and time required to carry out the study.

And, as an additional consideration, prevention controls that have been correlated with physical testing—either as part of research activities or previous programs—are likely to garner a smaller number than will a simulation or computation that has little or even no physical corroboration.

TECHNIQUES FOR CONTROL AND DETECTION ANALYSIS

To devise controls and rate detection both effectively and efficiently, you should again work column by column on the worksheet. This can be modified a bit, as you will see in a moment, but the best way is still a column-by-column approach.

To start, tackle each C-M-E chain separately. Then, consider what might be done to detect a cause before building any prototype hardware. This will constitute a prevention control. Next, reflect on what type of testing could be done using prototypes—in other words, brainstorm a detection control for this chain.

Of course, you should make an effort to brainstorm the most effective type of control for both causal factors and for effects. At a minimum, you must have one control for each chain. However, having both prevention controls and detection controls will, in most cases, provide the most effective way of verifying design issues.

Item/ function	Potential failure mode	Potential effect(s) of failure	Severity	Classification	Potential cause(s)/ mechanism(s) of failure	Occurrence	Current design controls Prevention	Current design controls Detection	Detection	RPN
Function from Step 2	Mode from Step 3	Effect from Step 4	X		Cause from Step 5	Y	Action against causal factor	Physical test for effect	Z	

Figure 8.3 The relationships between causes and prevention controls, and effects and detection controls.

What this really means is that each type of control is based on a different column in the worksheet. Detection controls are aimed at effects (developed in Step 4) while prevention controls are aimed at causes (developed in Step 5). On the DFMEA worksheet, this looks approximately like Figure 8.3.

As Figure 8.3 shows, prevention controls consist of actions that aid in discovering the legitimacy of causal factors, while detection controls are actions that ascertain the validity of failure effects.

After you have brainstormed both prevention controls and detection controls, you are then ready to use a detection table to assign a risk number to these controls, remembering to rate only the control that's most effective for each C-M-E chain.

And, once again, if you are debating about numbers when you rate detection, don't spend any time debating a difference of one number—use the pessimistic higher number. Just remember that you are justified in using a lower number when comparing prevention and detection controls. If you've done a sound job in the preceding steps, this should be a relatively quick and, in most cases, easy set of tasks.

SUMMING UP: STEP 6

Developing controls and rating detection is not hard, nor should it be time-consuming. The process is essentially the same as that employed in Steps 4 and 5, as Figure 8.4 shows.

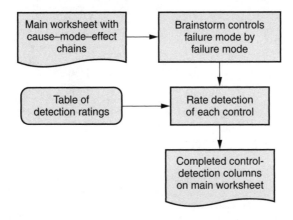

Figure 8.4 A summary of Step 6 actions.

After completing the controls and rating detections, the weldment project (with all the caveats we've previously discussed) could look like Figure 8.5. In this table, you will note that adding a second effect of "rear wheel traction is decreased" to the first effect of "steering precision is diminished" has no direct impact on the controls. The control for both effects is the same (prototype durability test). The detection rating is not the same, though, because prevention controls for these C-M-E chains are a bit different, and the rating is based on the most effective control. In addition, in each case the prevention control rating is better than the detection control rating.

Item/function	Functional requirement(s)	Potential failure mode	Potential effect(s) of failure	Severity	Classification	Potential cause(s)/mechanism(s) of failure	Occurrence	Current design controls Prevention	Current design controls Detection	Detection
Locate rod to bushing ring	14.5 mm from bushing face to rod OD	Rod partially located	Rod–ring interface fractures; loss of vehicle control	10		Angular relationship between rod and ring incomplete	2	Design review	Prototype component durability test	3
		Rod loses location over time	Steering precision is diminished	7		Rod section modulus insufficient	3	Finite element analysis	Prototype vehicle durability test	2
			Rear wheel traction is decreased	7		Incorrect weldment callout	5	Design review and comparison against historical data	Prototype vehicle durability test	3
		Rod incorrectly located	Premature tire wear	5		Incorrect suspension geometry provided by system engineering	2	Review of data with system engineering	Prototype vehicle durability test	3
						Angular relationship between rod and ring incorrect	3	Suspension geometry simulation	Prototype vehicle durability test	2

Figure 8.5   The DFMEA worksheet with controls and detection ratings completed.

9
Step 7—Assessing and Addressing Risk

W e've come a very long way in the DFMEA process. Now we can begin to systematically evaluate risk, which is one of the major reasons we've done all of the work up to this point (see Figure 9.1). To assess risk, we have to understand some very basic and, in my opinion, inevitable issues:

- No single indicator of risk is likely to provide the full and complete story.

- DFMEA, despite being a highly deductive process when done properly, is neither an ironclad nor perfect indicator of all risk factors. If you are looking for perfection, you must first live an exemplary life and then die; no system of belief in the world today professes the potential for perfection in this life.

- The legal system in which your business operates, particularly the civil (non-criminal) aspects of that system, will have an outsized impact on how risk is assessed. This means that technical evaluations of risk must be tempered with the impact of laws, legal opinion, and, in many cases, public opinion.

- Costs, which have only been an indirect factor in our discussion thus far, can not be ignored when discussing risk. As Dr. W. E.

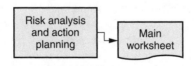

Figure 9.1 Step 7 of DFMEA process.

Deming noted, the most significant costs are often unknown and even unknowable.

• Management must be prepared to accept some level of risk with any and all products. A risk-free product is fantasy, and the extremely risk-averse dictates of some managers can easily cause the DFMEA process to collapse into a modern-day Kabuki dance.

RISK PRIORITY NUMBER

The *risk priority number,* or RPN, is easy to understand. It is simply the product of severity, occurrence, and detection. Mathematically, RPN = S × O × D, which can be remembered as "S-O-D." This is one indicator of risk and it has some significance, but it is not a magic number that determines what you should do to address risk.

The weakness of RPN as an indicator of risk is also easy to understand. RPN assumes coequal significance for severity, occurrence, and detection—and this makes RPN less analytical or prescriptive than many people recognize or would like. The reality is that severity has a greater impact on risk than occurrence, and occurrence is more significant for risk assessment than detection.

In other words, the results of a failure are the most important aspect of risk. How often a failure might occur is next in significance, and finally, your ability to detect the failure (presumably before the effect becomes manifest) is of lesser importance. After all, if the failure rarely arises, a large risk for non-detection pales.

For example, assume an S-O-D of 7-3-4. This would yield an RPN of 84. Would this be a greater or lesser risk than an S-O-D of 7-5-2, yielding an RPN of 70? Would you really be willing to bet the significant market risk of a severity of 7 on controls and detection? An occurrence of 5 is disturbing in most products, and even a detection of 2, despite being stringent, is far from flawless.

Personally, I'd prefer the 7-3-4 scenario to the 7-5-2 scenario in most cases. However, I would want to know a great deal more about the nature of the product, the marketplace usage of the product, and the specific issues that led to a severity rating of 7 before I decided what actions I would be willing to consider in addressing this situation.

Last but not least, I want to caution against the number one stupid, foolish, and shortsighted thing that can be done regarding RPN. In too many organizations, managers decide to set a threshold value for RPN that requires corrective action. The goal of making sure that all risk is below a certain level, while admirable, is impractical and unworkable.

To start with, the number one result of setting a threshold is that DFMEA teams will manipulate the S-O-D ratings if the RPN is close to the limiting value. For example, let's assume that a C-M-E chain, with the appropriate controls and detection rating, has an S-O-D of 9-3-3, for an RPN of 81. If the threshold is 75, I can almost assure you that an experienced DFMEA team will decide that either the occurrence or detection rating is really 2 instead of 3, yielding either a 9-2-3 or 9-3-2 pattern and producing an RPN of 54.

The result of this is again simple—and, perhaps, simple-minded. The team will pretend that there really isn't a meaningful risk, and management will pretend to believe it. Then, later, when this risk comes to the fore (perhaps—there's no certainty), management will be unhappy, even irate, that the real nature of the risk wasn't explained.

Numbers can be assigned, but simplistic judgments based on RPN will be valued and acted on in different ways. However, senior managers always seem to be looking for a single parameter (RPN, total cost, C_{pk}, whatever) that will allow an easy and quantitative assessment of risk, an assessment that requires minimal judgment on their behalf.

I believe such thinking is both naive and one-dimensional. Perhaps one day we will develop ideas that will move toward a fully quantitative risk assessment[1] based on a single calculated parameter. But the current reality is that management must fully understand what the situation is and make an informed decision about risks that DFMEA highlights.

To look at it another way, let's assume that the DFMEA team honestly and forthrightly reported an S-O-D of 9-3-3, again leading to an RPN of 81. However, the only two options that anyone can think of to address this are to either create a new test machine that costs twice the development budget of the project itself or to make a design change that would increase the product cost by 50%. Would either of these choices be financially sound?

It's doubtful that any management team would look at this and think this is a wise way to spend the shareholder's money. On the other hand, is management ready to live with the risk that's inherent in this scenario?

---

1. Some have proposed either a total cost assessment or a weighted S-O-D scale to address this. Total cost falls short since contingent costs, particularly when injury or death are potential effects, are extremely difficult to assess. Similarly, weighted S-O-D scales, which could compensate for the unequal nature of severity, occurrence, and detection, are flawed; each project seems to require a different weighting scale. And the scales are again notional or qualitative in nature, just as the S-O-D tables are. Ultimately, you end up piling guesstimate upon guesstimate, and the value of the information isn't worth much.

The nature of risk is just that—there's peril that must be faced. No design, no technology is free from the potential for negative outcomes. In the aerospace business, risk is inherent and significant. Managing this risk in any business presents numerous technical as well as financial challenges, but it is immature to think that risk can be completely and fully overcome.

This will be a recurring theme as we continue this discussion about risk.

CLASSIFICATION AND CHARACTERISTICS DESIGNATION

At the end of the day, risk is inevitable in any product development project. DFMEA provides important insights that help identify important or noteworthy risks and provides decision makers with detailed information about these risks. This will allow a project team to minimize (but not eliminate) the potential impact of risk.

It's also important to understand that classification and special characteristics play an important role in verification—as well as a critical part in validation *and* in ongoing manufacturing process control. There are dozens of callouts and specifications associated with even the simplest of products.

Are all of these subject to intense scrutiny on every prototype? Is every dimension or every performance criterion subjected to 100% inspection during serial or ongoing manufacturing? Of course not—the expenditure of time and money to do so is absurd in all but the most unique projects. Only the most important issues, those that are classified as special, merit the expenditure of time and money in verification, validation, and process control.

Classification

To do this, though, the team will need to identify the characteristics of the design that drive the underlying risk in a C-M-E chain. In DFMEA, this is all about product characteristics. In PFMEA we can have the added complexity of process parameters, but for DFMEA this is simply and singly about design characteristics.

In looking at each line of the DFMEA, an S-O-D pattern will be produced and will be linked to a C-M-E chain. If the severity–occurrence pattern, in particular, meets certain criteria, then the risk is very likely to be significant.

The first action needed in this subprocess is to decide if a "classification" indicator should be entered onto the worksheet. This can be based on a variety of issues, but classification is usually based on a concept called *criticality*. Criticality is an old idea; it was first introduced in the initial U.S. military specification for FMECA issued in 1949 that was discussed in Chapter 1.

In the simplest terms, criticality is based on the product of severity and occurrence. This is premised on the idea that severity is the most important aspect of risk while occurrence is less important but still more important than detection. After all, if something is unlikely to occur, a less-than-ideal ability to detect the issue really isn't that important.

There are many schemes for using criticality and determining a classification for any C-M-E chain in DFMEA. Most use two categories for classification:

A *critical classification* (sometimes called "CC") results whenever there is a realistic possibility of injury, death, noncompliance with a significant government regulation or statutory law that may result from a function disruption.

A *significant classification* (or "SC") results whenever a major feature or function of the product is likely to be disrupted in a way that is both readily apparent to customers and will likely cause noteworthy dissatisfaction.

Clearly, "critical" has greater importance than "significant," but what indicators—taken from the work we have done thus far—can be used to decide which chains are critical and which are significant?

The first and perhaps least sophisticated approach is to set an arbitrary value of criticality that results in a critical classification, followed by a lesser but still arbitrary value of criticality that results in a significant classification. This, though, has two weaknesses. Setting arbitrary limits often results in management of the ratings, just as it does for an arbitrary RPN criterion. In addition, this still doesn't reflect the fact that severity is more important than occurrence, and therefore a very high occurrence combined with a lower severity can have the same criticality as a high-severity, low-occurrence C-M-E chain. Clearly, these two situations are not equivalent.

The second, somewhat more sophisticated approach is to calculate all of the criticality values for C-M-E chains and then perform a Pareto analysis on the resulting tabulation. Those in the highest echelon gain a critical classification, those in the secondary echelon are designated as significant, and all remaining chains (still potentially important but clearly less so) are given no designation.

The weakness of this approach is that it ignores the fact that the severity scale has a high degree of nonlinearity at the top end. A severity of 9 or 10 may have much greater import than a severity of 8, and yet the difference in rating value is small. And this still does not address the concern that severity is more noteworthy than occurrence.

Finally, the most common and least arbitrary method is to devise a simple set of logic rules that differentiate between S-O patterns. This recognizes the nonlinearity of severity and also gives severity greater weight in the assignment of classification. This approach, which was pioneered at Ford, uses two relatively simple rules:

1. A *critical* classification will result whenever there is a severity of 9 or 10 (or whatever value on the severity scale indicates the potential for injury, death, regulatory failure, or violation of statutory law). Occurrence and detection ratings are irrelevant; the classification of critical is assigned whenever effects exhibit a sufficiently high severity.

2. A *significant* classification will result whenever a lesser severity is anticipated and the occurrence for this C-M-E chain is 4 or higher. In general, most tables use severity ratings of 5 through 8 to indicate that the effect is both evident to customers and will lead to a meaningful level of discontent, a good range of outcomes for something that could be significant.

Any of these schemes can be used. The scheme that you select should be based on your organization's understanding of these issues. Some organizations use graphic symbols or icons instead of "CC" and "SC," but the principles are almost always similar to those outlined above.

Identification and Designation of Characteristics

Any time there is a classification of *critical* or *significant* for a C-M-E chain, one or more product characteristics—the specifications that were discussed near the end of Chapter 4—will require a special designation. For each C-M-E chain that is either critical or significant, one or more characteristics must be identified as critical or significant.

The challenge in this action is simple. The C-M-E chain doesn't necessarily identify the specific characteristic that needs special identification. On the DFMEA worksheet, you are classifying the cause–mode–effect chain, not particular characteristics of the design. That's a separate activity that we will address in Chapter 10.

We do need to discuss this situation, though, because many DFMEA practitioners have difficulty separating the classification of a C-M-E chain from the designation of a critical or significant characteristic. In many, but certainly not all, cases the characteristic that must be identified is the issue spelled out in the second column on the worksheet, namely the functional specification for the function under consideration.

However, it's often true that the specification callout in the second column isn't detailed enough to offer the information that's needed to properly manage risk. This is particularly true for system and subsystem analyses[2] but can even arise in the most mundane component-level analysis. For example, in the weldment example, we've been looking at a specification of "locate rod to bushing ring 14.5 mm from bushing face to rod OD," a relatively simple mechanical condition.

Does this communicate the proper level of concern? Is this the characteristic that requires the most attention, both in the chain of verification and in the chain of validation? This isn't the easiest dimension to measure. Moreover, it might not be the most indicative dimension, either. It may be easier and more useful to measure the length of the bushing or to measure the outer diameter of the rod or some other characteristic.

In some cases, a function will have a complex performance specification that can be particularly difficult to measure, such as a mechanical yield point or complex impedances in a circuit. Some other characteristic, say the hardness of the material or the reactance in just a portion of a circuit, might be more appropriate as a designated characteristic.

It can also be that one characteristic addresses multiple classified C-M-E chains; it may also be true that one classified C-M-E chain might possess multiple characteristics that must be considered critical or significant.

At this stage, however, all that is necessary is to enter the correct classification for each relevant C-M-E chain in the "Classification" column. You can use whatever scheme is appropriate, but enter the appropriate abbreviation or symbol in this column. We'll return to characteristics designation in Chapter 10.

With classification and RPN entered on the worksheet, you've effectively analyzed the existing concept design. One final note of caution is worthwhile, though. Too often, teams put too much emphasis on the values for S-O-D and RPN. At various points in this book, I've tried to explain the factors and motivations that drive this peculiar mania.

---

2. This would be indicated by the location of the column-segment in the block diagram.

When all is said and done, the numbers you have entered on the worksheet are just that—numbers. They're not even the most important numbers, and repeatedly I've urged a conservative approach (when in doubt, use the higher number) to assessing risk. After you've completed the chain of verification, you will have a chance to revisit all of the numbers; you can see on the far right side of the DFMEA worksheet where these revised numbers can be entered. We'll discuss that in more detail in the next chapter, too.

You should also now be able to understand why we skipped the classification column earlier in the process. Classification is most dependent on severity, but you also need occurrence to really assess classification, and the general practice is to address this at the same time that RPN evaluation and action planning are taking place.

We now turn our attention to potential design improvements or actions that can be taken to reduce risk.

RECOMMENDED ACTIONS

Risk factors are not ordained in the heavens. In DFMEA, by altering one or more design choices, it is possible to change the level of risk in a design. This can be done primarily by altering design factors that affect occurrence or, to a lesser degree, by altering one or both of the controls for any C-M-E chain. However, as we will see in just a minute, altering severities is usually either too difficult or too expensive to be practical.

Nonetheless, at this point a project team needs to decide what, if anything, should or can be done to reduce the risk associated with the C-M-E chains found on the DFMEA. While the details of any change that can be made will differ for each type of product, and probably each individual project, there are some general principles or strategies that can be applied to address risk factors in design.

Reducing Severity

It's extremely difficult to reduce severity. As we saw in Chapter 6, once a function is disrupted it's not really possible to know which effect might arise. And, if one of the reasonably possible effects has a relatively high severity, that outcome is certainly possible, and accepting the risk associated with that severity must be considered a realistic prospect.

In practice, severity reduction is barely possible, with one largely and nearly always impractical exception. For most engineers, conceiving and executing a significant reduction in severity that is cost-effective and doesn't transfer risk elsewhere is a once-in-a-career event. However, for reference, here are some possible ways that severity might (just might) be reduced:[3]

- Eliminate the failure cause through a design change. This is easy to say but extremely difficult to do. And, in many cases, if you manage to do this, you will simply move the risk to another C-M-E chain.

- Decouple cause and effect—in other words, change the design to make the effect impossible if the underlying function is disrupted.

- Constrain usage through system-level lockouts or fail-safe design considerations—a complex activity that often involves many other organizations, costs a great deal, and is similar to the redundancy concept discussed below.

- Change an existing design feature so that failure is less dramatic or catastrophic. In practice, I've never actually seen this done (unless by redundancy, which, again, is a special case). I believe it's achievable, though, so I've included it as a possibility.

- Finally, the most common way to reduce severity is to increase redundancy in a design.

Redundancy virtually always adds significant, even prohibitive, cost to a design and, in most cases, this is simply not feasible. In addition, redundancy simply transfers the risk to another C-M-E chain. While the new C-M-E chain will very likely have a lower occurrence, the severity of the effect will, in nearly every circumstance, remain high.

In the auto industry, two common design features in modern vehicles illustrate these principles:

- Why does a car have two headlights? Simple—if one headlight assembly fails to illuminate the road, the other will still provide partial illumination, reducing the severity of a bulb failure. Nevertheless, the cost to have two headlamp assemblies is very significant—with today's optics and lighting capability, a single lamp in the center of the vehicle could adequately illuminate the road—yet no vehicle company seems willing to take the risk that

---

3. I can't even provide examples for these principles; the very few times that I've seen one of these instances, they've become an important proprietary feature in a product, and that particular intellectual property is protected by confidentiality agreements that I can not abrogate.

would arise with only one lamp assembly.[4] And the risk of total failure will still exist in the electrical circuitry that feeds current to a single lamp assembly.

- Airbags protect against the failure of a seat belt—but add significant cost and have dozens of C-M-E chains with effects every bit as severe as the effects associated with seat belt failure modes. Moreover, the C-M-E chains that make seat belt designs risky still don't go away because there is always a possibility that an airbag won't deploy, will deploy improperly, or will still cause injury if a seat belt fails in a crash.

On the other hand, aerospace designs often use redundancy because the severity of some effects is simply too awful to allow. Many airplanes have triple-redundant hydraulic systems, because the failure of a main pump or a critical hose would almost certainly lead to a crash if there were no backup system. This isn't inexpensive, though, and in most circumstances the cost would be prohibitive to have that kind of belt-and-suspenders design.

Actions that reduce severity nearly always increase cost and/or transfer the risk to some other feature or system in a product. As a result, it's very difficult, verging on impossible, to reduce severity in most cases.

Despite the fact that some engineers (and even more managers) find this either disconcerting or objectionable, the fact remains: reducing severity is extremely difficult. Even the effort to develop a solution that reduces severity is rarely worth the time and money needed to achieve this elusive end.[5]

Reducing Occurrence

Changing a design to reduce occurrence is perhaps the most powerful and straightforward action that can be taken to reduce risk. The goal is simple: make the cause—not the mode or the effect—less likely to arise. For example, the use of composite materials for structural loading often has a high occurrence because the range of strength for most complex composite shapes is less certain than for other materials. Changing to a better-understood material will, in most cases, reduce the occurrence of the

---

4. You can see that snowmobiles, ATVs, and even personal watercraft are adopting a two-headlight approach.

5. It might be possible to change a severity of 10 to a 9 with a modest change that provided some level of warning to the user of the product. This may, in fact, be valuable—but it really doesn't change the serious level of risk connected with injury or death that is usually associated with a 9 or 10 severity rating in most common rating scales.

causal factor itself. That may cause other trade-offs, as almost any major design change usually does, but it would reduce occurrence for the mechanical loading issues that were under consideration.

Alternatively, you can do something to make the cause less likely to lead to the specific failure mode in question. This can be done by increasing (or decreasing, as appropriate), the median value of a property so that the "tail" of the property distribution is no longer in a state or at a value that could result in a disruption of failure.

So, if you still wanted to use a complex composite, you might choose a higher-strength material. In essence, you are simply adding a factor of safety to the system when you do this, and this kind of change does reduce occurrence.

Finally, if the failure mode is driven by a loss of function over time, anything that can be done to increase durability or "life" of the product can effectively reduce occurrence.

It is often true, though, that reducing occurrence increases cost. This is usually less than the cost increases associated with severity reduction, but it is a reality that you must consider. In some cases, cost increases can be reduced by adding system-level features that will be visible upstream in the block diagram.

In addition, reducing occurrence will, at times, cause a change in some other associated property that causes difficulty. In the composite example, changing to magnesium could potentially create fire risks or lead to properties that increase the occurrence of other causal factors in separate C-M-E chains. Using design of experiments techniques can, in some cases, provide insight into complex changes that can be made that will satisfy multiple criteria with minimal cost.

Overall, reducing occurrence is usually not a simple and easy thing to do—but it's almost always the easiest thing you can do to reduce risk.

Improving Detection

Reducing the risk associated with controls can be relatively uncomplicated, but again it can add cost to a project. However, changing a control in a way that reduces the detection rating almost never changes another C-M-E chain, so collateral risks are rarely affected by this kind of action.

It is true, though, that changing detection can not change a C-M-E classification because detection has no effect on the designation of either a *critical* or *significant* classification in most classification schemes.

There are many things that can be done to reduce a detection rating. The first and most powerful is to detect causes using analysis rather than relying on testing. In other words, prevention controls are typically more effective at discovering causal factors than testing is for discovering effects (and then backtracking to discover causal factors). This occurs because the discovery occurs earlier in the chain of verification—and because good simulation or calculation can reasonably evaluate limit-of-property conditions.

All in all, the use of mathematical reliability methods, a subcategory of simulation and calculation, is one of the most powerful tools that can be applied to any C-M-E chain. Because this is often tedious and requires special expertise, it's not often used. But it does work well when done properly.

If a big detection rating results from less-than-effective testing, there are many different things that can be done. Here is a partial list of the most common things that can be done to improve testing and reduce detection ratings that apply to detection controls:

- Change testing procedures to more closely replicate real-world usage.

- Alter test evaluation criteria to provide additional insight and understanding of test outcomes.

- Increase failure feedback. In practice, this means testing to failure rather than testing to a limit and then terminating the test. The predictive power of Weibull analysis (or other failure distribution models) is much better than the predictive power of "bogey" tests that simply run to a predetermined limit and then stop if nothing untoward happens. However, because most test lab supervisors want to run a fixed schedule for testing, it's more common to test to a limit and then stop—despite the fact that this generates far less information and has a greatly reduced statistical importance.

- Increase the number of samples tested.

- Change testing instrumentation to create more-powerful understanding of what occurs during a test.

ACTION PLANNING

Through a great deal of work, you've now gained a very sound understanding of the concept design. The next set of tasks on the agenda is to make use of this information, again using systematic and, whenever possible, deductive methods.

To get the most from a DFMEA study, you should now apply this information in the chain of verification. This consists of the following sequential activities:

1. Using the controls developed in Step 6, create a comprehensive design verification (DV) plan. Make sure that you include both the prevention controls and detection controls entered on the DFMEA worksheet.[6]

2. Generate detailed drawings and specifications for the entire project.

3. Carry out the prevention controls from the DV plan using fully detailed design information. Make changes to the design if any of the prevention controls reveal flaws or weaknesses.

4. Construct prototypes, using appropriate small-quantity quality assessments, based on the corrected design data from the previous activity.

5. Complete the physical tests called out in the DV plan. Make changes to the design as required, update the DFMEA worksheet, and prepare all engineering documentation and records for release.

6. Record the results from all control activities as the final design verification plan and report, and obtain final approvals for all engineering documents.

7. Release the design for manufacture (or implementation for software or service projects).

While all of these actions are important, the most important is the second: the creation of detailed design activities. To prepare for this, you need to complete the next two columns on the DFMEA worksheet—the "Recommended Actions" and the "Responsibility and Target Completion Date" columns.

To start, you'll have to make some judgments. These decisions will bring to bear everything you've learned about your project and all of the principles of risk management that have been discussed in this chapter.

Each C-M-E chain presents a potential for improvement. However, in any real project, you are unlikely to have the time or resources to attempt meaningful enhancement for each and every chain. Moreover, to do so would probably be foolish. No design is ever perfect, and some degree of risk will almost always be present. So, what can you do?

---

6. Bear in mind that a solid DV plan is more than a list of evaluations and tests; it describes specific procedures, number of iterations in a simulation or number of samples tested, the conditions of testing, assumptions in calculations, specific criteria for comparison against standards, and many other factors that are too numerous to list in the DFMEA itself.

First, you should consider C-M-E chains with the highest RPNs as candidates for some form of remedial engineering. You should also consider C-M-E chains that are classified as either *critical* or *significant,* even if the RPN values for these chains are not high.

Having said that, however, it's often true that *critical* or *significant* C-M-E chains often can't be changed much. A *critical* chain with an RPN value of 40 could arise from a 10-2-2 S-O-D pattern. Can much be done about this? The severity is what it is and probably can't be changed. The occurrence is probably as low as it can be, and detection is a small number.

Perhaps the occurrence or detection could be lowered by one number, changing the RPN value to 20. If this could be done for a very limited cost, then it might be worthwhile. In most cases, though, an RPN value of 40, given a severity of 10, is about as low as can be expected. That's just the level of risk that is inherent in a product that has severities of 9 or 10.

By looking at these candidate chains, you need to decide which chains are, in essence, *ugly.* What does this mean? A chain is ugly when an occurrence or detection rating is higher than it ought to be and really needs to be addressed—or, in more direct terms, when the risk is higher than you can tolerate.

At the same time, your ability to address every single ugly chain, particularly on a complex project, is limited. So, you will have to decide which chains are so ugly that you need to do something.

For each ugly chain that can't be tolerated, you should develop some form of corrective action that will address either the occurrence rating or the detection rating. Lowering occurrence will alleviate ugliness more effectively than lowering detection. In any event, you need to apply the strategies discussed previously and apply them to the chains that are too ugly to accept.

On the worksheet, you need to summarize the bare outline of a miniature project plan. That means a brief summary of the actions you plan to take to address the ugliness (the what of the plan), as well as the person responsible for making sure this work gets done (the who of the plan) and the target completion date (the when of the plan).

Of course, you may very well need to work out a much more detailed project plan after you've described the essential elements on the DFMEA worksheet. If that's the case, that's another item for your "to-do" list, and you should use the normal tools of project management to plan and execute any recommended or corrective action that is more complicated than a simple one-step action.

One additional remark is advisable about these two columns. If you don't plan to carry out any action for a C-M-E chain, enter "None" in the

"Recommended Actions" column. This ensures that anyone reviewing the DFMEA worksheet can clearly see that you've consciously considered that C-M-E chain and decided to do nothing. If you don't do this, it might be perceived that you either just missed a chain or you didn't complete a plan for a chain. Either of these actions have very negative implications, as you will be able to see when we discuss product liability considerations in Chapter 10.

And—a final major caution—if you enter an action plan, make sure you complete it. Even if the planned actions are unsuccessful and have no impact on risk, *carry out the planned actions*. If you fail to follow up on an action plan and the underlying risk then appears as a defective product or worse, people will get angry.

They will get very angry. "You mean you knew this was a problem and didn't do anything about it? You even had a plan to fix this and you didn't do it?" You can hear the recriminations fly whenever this happens.

So, look at candidate chains and make sure that you have a realistic set of action plans. Don't promise more than you can deliver, but don't let any truly ugly C-M-E chains go without action, either.

SUMMING UP: STEP 7

Most of Step 7 is simple and reasonably easy to understand. You calculate the risk priority number, assign classifications as appropriate, and formulate action plans (see Figure 9.2).

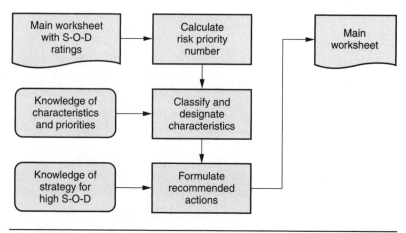

Figure 9.2 A summary of Step 7 activities.

After completing the controls and rating detections, the weldment project could look like the table shown in Figure 9.3.

You'll likely notice that I haven't included details about responsibility and target completion date. That would be very specific to an actual organization and project and doesn't really add anything to your understanding of the DFMEA process.

Item/function	Functional requirement(s)	Potential failure mode	Potential effect(s) of failure	Severity	Classification	Potential cause(s)/mechanism(s) of failure	Occurrence	Current design controls Prevention	Current design controls Detection	Detection	RPN	Recommended action(s)
Locate rod to bushing ring	14.5 mm from bushing face to rod OD	Rod partially located	Rod–ring interface fractures; loss of vehicle control	10	CC	Angular relationship between rod and ring incomplete	2	Design review	Prototype component durability test	3	60	Review of dimensioning for similar designs; FEA/fracture mechanics study at minimum conditions
			Steering precision is diminished	7		Rod section modulus insufficient	3	Finite element analysis	Prototype vehicle durability test	2	42	Change rod section if FEA suggests a problem
		Rod loses location over time	Rear wheel traction is decreased	7	SC	Incorrect weldment callout	5	Design review and comparison against historical data	Prototype vehicle durability test	3	105	Review of American Welding Society standard D8.8M:2007
		Rod incorrectly located	Premature tire wear	5		Incorrect suspension geometry provided by system engineering	2	Review of data with system engineering	Prototype vehicle test	4	40	None
						Angular relationship between rod and ring incorrect	3	Suspension geometry simulation	Prototype vehicle test	2	30	None

Figure 9.3   The DFMEA worksheet with selected recommended actions.

10

Using DFMEA
Constructively

If you've followed the process that's been described so far, you've realized that a sound DFMEA is a major undertaking.

How can you make this worthwhile—and not just a "check the box" activity?

A well-conceived and executed DFMEA is one of the most important records in an organization. If it's maintained properly and used constructively, it can provide a decade or more of keen understanding that can reduce problems, improve margins, and speed development cycles. It can also prevent major fiascos that can result from design missteps—flaws like the use of pure oxygen in the Apollo program—that can cost lives and huge sums of money.

None of these outcomes, though, are automatic. As is almost always the case in product development, you must carry out serious work to get real benefits. In this final chapter, we'll look some of the most important follow-up activities that are needed to reap the rewards of a first-rate DFMEA study.

Beyond specific follow-up activities, we'll also discuss some of the legal issues that often arise whenever an organization begins to fully understand the benefits and limitations of deductive DFMEA studies. We'll also discuss how DFMEA activities affect organizational quality systems in general and how the chains of verification and validation are linked by DFMEA information.

SPECIAL CHARACTERISTIC LISTINGS

In Step 7, we saw how cause–mode–effect chains can be classified for risk. We also discussed how each classified chain will require the identification of specific product characteristics on engineering drawings.

That's something that needs to be done during the detail design stage of the chain of verification. A comprehensive listing of all critical and significant characteristics (or whatever terms your organization or business sector uses for these concepts) is an essential element of any decent quality system.

Let's review the ideas that compel characteristic identification (as opposed to chain classification) in a sound quality system:

- Each classified C-M-E chain must be associated with at least one special product characteristic that drives risk.

- Characteristics that drive risk for classified C-M-E chains are often called *designated characteristics* or *special characteristics*. They are also frequently called *critical* or *significant characteristics*.

- Designated or special characteristics are not only important in the chain of verification but also play a central role in the chain of validation as well as ongoing process control.

- Regardless of the terminology employed, formal identification or designation (on engineering documents) of one or more characteristics that drive risk for every classified C-M-E chain is extremely important.

  - A classified chain may be driven by a single product characteristic or by several product characteristics.

  - Similarly, one product characteristic can drive risk for many classified chains.

  - In other words, the relationship between a classified chain and product characteristics can be one-to-one, one-to-many, or many-to-one.

  - Many-to-many relationships probably exist for a lot of products, but tracing such complex relationships is really beyond the scope or capability of the DFMEA process. Don't bother pursuing this level of understanding unless you are prepared to embark on very sophisticated analyses to make use of this information.

- Characteristics that drive risk for a classified C-M-E chain must be based on the functional specification entered in the second column of the DFMEA worksheet.

  - In some cases, the functional specification described in the worksheet that the classified C-M-E chain is built upon will be adequate and can be designated as a special characteristic that drives risk.

– In other cases, the specification described in the worksheet may require further breakdown into several different detailed characteristics. When this occurs, one, two, or all of these subordinate characteristics may be the driver or drivers of risk, and one or all of these characteristics could well merit designation as special.

- Designated characteristics must be identified not only on engineering drawings, but on specifications, purchase orders (when subcontractors are involved), and on many different quality system documents and records, including process control plans. Identification can be by letter, symbol, or other appropriate "callout" methods.

– This is particularly important in achieving a sound linkage between verification and validation activities. This means that production-level input—always an important element in a DFMEA team—must be given full weight when designating special characteristics in DFMEA.

When you first attempt to do this, you may find that you are confused about the difference between a classified C-M-E chain and a designated characteristic. If that occurs, step back and review the ideas listed above.

In the end, you will find that designating characteristics requires engineering or technical judgment. The DFMEA process will make this a largely deductive process, but real engineering is too complex and difficult to reduce to a set of equations, a matrix, or a syllogism.

You simply must use your mind—and your judgment—to be a good product engineer. The flow diagram shown in Figure 10.1 recaps the major steps you should follow to accomplish this.

DESIGNATED CHARACTERISTICS IN VALIDATION AND PROCESS CONTROL

The information that's developed in a solid DFMEA study can and should be used for overall quality improvement, not just for better design and verification activities. In particular, designated characteristics should form a major part of the basis for improved validation of production as well as continuing process control.

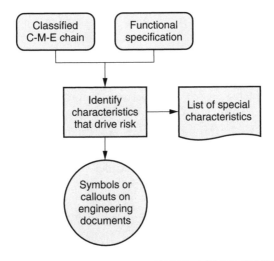

Figure 10.1 Designating characteristics using classified C-M-E chains.

The chain of validation and ongoing process control are focused on several issues, but by far the most important is delivering consistent, problem-free products to customers. And by problem-free, I mean products that are both pleasing and free of unpleasant surprises. Of course, PFMEA certainly addresses that issue, but everyone needs to realize that DFMEA is a major input to PFMEA.

This relationship occurs for two reasons. To start, DFMEA should be, as discussed in Chapter 2, a team effort, and that team should include representation from the production side of the business. Also, production is really the way that the design moves from a theoretical state to a state of reality. A design is a complex set of information that the production staff needs to translate into some form of tangible product that customers can see, touch, and perhaps hear, taste, and smell, as well.

If you understand that, you will realize that the risks associated with design are, unavoidably, similar to risks associated with production. All that the production process can realistically expect to accomplish is to create and deliver products that faithfully execute the requirements spelled out in design information. The interaction of risk management between design and production is particularly important for effects and severity, and for designated characteristics.[1]

---

1. Of course, designated characteristics are primarily driven by severity ratings in classified C-M-E chains.

In the chain of validation, PFMEA plays the same pivotal role that DFMEA plays in the chain of verification. This means that risk assessment—with a focus on customer perception—is at the heart of PFMEA studies. More to the point, C-M-E chains in production will produce the same effects as those seen in DFMEA studies. Causes and controls will almost certainly be different, but the customer-centric effect of a production failure will be the same as the customer-centric effect of a design flaw.

After all, customers don't care, and rarely know, whether a defective functional element of a product is due to a design flaw that was accurately executed by a production group or whether it arises from a production error that mars a flawless design.

This means several things:

- Every major effect, specifically those that result in classified C-M-E chains, that is present in a DFMEA will very likely be present in one or more supporting PFMEA studies.

- The severity ratings of effects in PFMEA studies must be consistent with the severities found in DFMEA studies.

- Every designated characteristic from DFMEA requires specific attention in ongoing process control, whether that occurs in the build-up of a final assembly, a subassembly process, or a fabrication process. And these principles apply throughout any supply chain elements that support the final assembly of a product.

This final linkage—of designated characteristics from DFMEA to process control—often causes difficulty and is worth some discussion. Too often, DFMEA teams make a concerted effort (either conscious or otherwise) to limit the number of classified C-M-E chains because not limiting them is seen as forcing an "impractical" level of process control in production.

To avoid this, production planners, manufacturing engineers, and process control specialists (often the same people) need to realize that a product issue that can hurt someone or lead to serious customer dissatisfaction can not be ignored. On the other hand, "specific attention" need not be onerous in the production effort—it needs to be intelligent.

For example, more than a few organizations have settled on a simple rule: every critical characteristic requires statistical process control. That kind of rule, like the adoption of threshold values for RPN, is yet another attempt to turn risk management into a simplistic yes-or-no exercise that requires little if any consideration beyond compliance with an uncomplicated decree that can be managed with minimal effort.

Adopting that kind of approach will almost certainly degrade quality and profitability in an organization.

Let's take a look at a simple mechanical situation that is common in DFMEA yet often causes angst for production personnel. Most mechanical devices support loads. Whether these loads are bending moments or torques, compressive, tensile, or shear in nature, dynamic or static, loading is a common issue in mechanical systems.

The underlying function could be "resist deflection," "prevent deformation," or "limit fatigue," but there will be one or more functional aspects of a mechanical design that flow from loading in the DFMEA. And, more often than not, this may well lead to one classified C-M-E chain or even to several such chains.

What will be the driving characteristic in each case? It could be a cross-sectional profile or section modulus—or simply one dimension on the profile. It could be a material property, such as yield strength, or it could be a mechanical endurance limit.

Now let's say that the product is composed of several stampings that are assembled in some way—they might be welded, or riveted, or screwed together (it doesn't matter for this discussion; those features will be addressed in other C-M-E chains). And, after due consideration, the team has decided that the critical yield point on steel is the characteristic that drives several different classified C-M-E chains.

Does that really mean that statistical analysis of tensile test samples from every coil of steel should be used to control the manufacturing process? For anything but the most critical application, say, something that could kill thousands from a single defective product, conducting that much testing would be absurd or worse.

Specific attention for this characteristic can simply consist of requiring the steel supplier to provide mill certifications for composition and strength with every coil. Or, if the nature of the effect were a bit less severe, then you might well only need certification for a single heat rather than every coil of steel.

Once again, teams working on product development processes need to engage their brains before reacting. If you make intelligent yet practical decisions, pointed debate, argument, and incongruity about the translation of effects and severity in DFMEA into PFMEA and process control are, in the main, red herrings. They add nothing, take too much time, and ultimately cause degradation of validation and process control activities.

RECORDING ACTION PLAN RESULTS

While this is really an integral part of a DFMEA study, I've pulled this out of the main process flow because it often transpires long after the initial study is completed. But, once you've completed all of the recommended actions developed in Step 7, you need to record the results of your actions on the worksheet.

By recording the consequences of your efforts, you leave a firm record of what has been done. Even if a particular action was unsuccessful, you provide important information that can be used to understand risk, to guide future projects, and to reduce product liability exposure.

It's not necessary, though, to enter all of the details of your endeavors on the worksheet. Instead, you need to enter nothing more than a brief summary. Then, if there are additional details, say a report, a memo, photos, or other type of record that's been generated, you can either attach that record to the DFMEA or cross-reference the supporting record.[2]

Of course, it's also possible that you complete one or more action plans and you are not satisfied with the results. Then, you need to devise another action plan and add this second plan to the "Recommended Actions" column on the form. Don't remove the first plan, but add the second to it. Then, after you've completed this work, you should add the outcome of this second plan as well as the outcome of the first plan. Adding rather than replacing shows the history of what you attempted to do and leaves a record of what did and didn't work in the development of the project.

You could do this several times for a given issue, but the odds are that you won't repeat the action planning activities more than once. Even one repeat will be the exception rather than the rule for most corrective actions.

Still, in all, good records are worth a great deal. As George Santayana, the famed philosopher, poet, and novelist said, "Those who cannot remember the past are condemned to repeat it."

Make sure that your efforts pay dividends. Keep records of your actions to reduce risk.

LEGAL AND ETHICAL ISSUES

Too often, I encounter engineers who believe they must craft any and all information on a DFMEA worksheet in a way that will make the product look perfect. They do this because they have a largely wrong-headed idea that any admission of risk in the DFMEA will mean that their company can be sued for product liability concerns.

---

2. Quality system requirements in most organizations will dictate how and where records of quality activities are stored.

This is a thorny subject, and it's really not possible to address this fully without asking you to read another book—or two or three books, for that matter. In fact, you might have to study for a few years to really understand all the aspects of product liability.

Nevertheless, we need to spend a bit of time understanding both the legal and related ethical ramifications of DFMEA. In this discussion, I will focus on issues that are driven by the U.S. legal system. If your company does no business and has no presence in the United States, then most of this discussion regarding product liability issues probably won't apply to your work.

However, most businesses that would be interested in and regularly use DFMEA are either firmly entrenched in the U.S. business community or have a strategic goal to become active in the United States. Even if your products are made overseas and sold overseas, if your organization has any financial presence in the United States, you could become enmeshed in the legal mire that passes for product liability law in this country.

Before we begin, though, I must make a disclaimer. I'm not a lawyer, nor do I intend this as a comprehensive review of U.S. product liability law. If you have any doubts about product liability issues, seek counsel from a qualified lawyer.

The first point that must be made is that product liability is about *torts*. Torts are not criminal acts that can cause someone to go to jail (at least not directly). Instead, a tort is a civil act that may cause damage to another party. Furthermore, a tort is a non-criminal (or civil) act that is *not* a breach of contract. A tort is possible even if all aspects of a contractual obligation have been upheld.

Ultimately, a tort is an act for which a court may provide a remedy in the form of damages or an order for payment of money from one party to another. The idea is that the damaged party is compensated for losses that arose from the tort. To obtain relief—legal jargon for compensation—a damaged party must file and then win a lawsuit or an action against another party.

Torts can arise from willful acts as well as from acts of negligence. When products are sold to others, though, a tort can also arise from something called *strict liability,* which is neither willful nor negligent. To understand strict liability, we need to have another brief history review.

Lawsuits that are associated with product liability are usually based on common law rather than statutory law. This means that the outcome of a lawsuit involving a tort will be largely decided on the precedents set in previous lawsuits rather than some written law. While more than a few laws governing product liability have been enacted in the United States in the past few decades, most of the law regarding product liability is still based on common law precedents.

When it comes to products, a good deal of U.S. product liability law is based on British law. For a long time, the principle of *caveat emptor,* or "let the buyer beware," was the foundation of commerce in Western civilization—the concept arose in the Roman Empire and became a basic principle in British law. This meant that the buyer was responsible for anything that occurred with a product after it was purchased. This was carried over to the United States from British law as part of the legal system that developed after the American Revolution. This is certainly the oldest precedent in product liability.

The essential fairness of this proposition was based on a straightforward idea. Products were simple, and any buyer should make an effort to understand what was being bought and use it appropriately, and, if anything went wrong, the maker of the product could not be held responsible or accountable for the consequences unless the seller or maker made a deliberate effort to harm the buyer through the product.

As technology developed, the equilibrium assumed in *caveat emptor* between buyer and seller became unbalanced. Products became so complicated that buyers had no realistic hope of understanding all of the issues involved, and things began to change.

As an illustration, I have personally facilitated a number of DFMEA and PFMEA studies on airbags and seat belts. I'm also savvy about engineering in general (at least I think so). However, when I buy a car, I have no real in-depth understanding of what the restraint system in the car might or might not do. Since I could be killed as a result of something that the system does or doesn't do, *even if there's no crash,* this isn't trivial. Yet I couldn't really evaluate the safety of a given restraint system even if I tore the system apart and studied it for years.

I think I could understand a clay pot, though.

As this imbalance became more apparent, lawsuits from consumers or customers against manufacturing companies grew in number and complexity. The argument of disparity between seller and buyer was accepted in several instances in the United States, and the idea of "strict liability" soon became a key issue in product liability.[3]

Strict liability isn't based on what a manufacturer might or might not do to design and produce something. In the starkest interpretation of strict

---

3. At this point, U.S. and British product liability law sharply diverged; this happened over several years in the mid twentieth century.

liability, a manufacturer is liable for damages or compensation when a buyer is injured or suffers financial loss from using a product. It doesn't matter that a company used great care in designing and making the product; if the product has any kind of defect and the result is harm to the buyer, the manufacturer may (may!) be held liable to pay for that harm.

Over the years, strict liability has led to more than a few famous (and perhaps outrageous) lawsuits and outcomes, such as the legendary suit against the McDonald's restaurant chain filed by a woman who was severely burned when she spilled hot coffee onto her lap. While the facts in this lawsuit are often exaggerated or distorted, the woman who was burned was far from the first to be burned from hot coffee at the Golden Arches, and McDonald's was not unaware of the fact that this was not a completely isolated incident.

Eventually, McDonald's was found to be liable for damages, and the woman who was burned was awarded a significant sum of money. Through several appeals, the amount (almost $2.9 million in the initial judgment) was reduced to $400,000, and eventually McDonald's agreed to a secret settlement—so we really don't know how much it finally cost the "Mickey D's" folks to settle this.

No matter, McDonald's did pay a large sum to the burned woman, and two other things are now part of world culture as a result. The serving temperature of McDonald's coffee has been reduced, and all coffee cups at McDonald's have written warnings about possible burns.

Manufacturers are not defenseless in this arena, though. Strict liability isn't absolute. There are many different issues—involving due care of the manufacturer, warnings, well-known hazards, clear and improper usage of a product, use of a product while impaired or intoxicated—and other factors that can cause a given case to be judged against the buyer and for the producer.

However, there's one more factor that one should not forget about in product liability. In general, corporations—faceless bureaucracies, often with a limited public persona or reputation—are often thought of as guilty until proven innocent in product liability suits. That's not the legal basis that courts are built on in the United States, but things like Ford's difficulty with the Pinto and the Explorer have reinforced this, and it can't be discounted in understanding American legal practices in product liability.

Even when a producer wins a lawsuit, the company can easily lose in the so-called court of public opinion.

In these two situations involving Ford, the number of lawsuits was large. Many people suffered serious injuries, and there were dozens of deaths. In other situations, product quality problems have evolved into class actions, where many damaged parties combine their legal efforts (and their staying power in long court proceedings) to win damages.[4]

Notwithstanding company victories in product liability lawsuits, the damage to public perceptions about the company's products can be terrible. Even today, almost forty years after the fact, the Pinto fuel tank problem hasn't been forgotten in some parts of the marketplace.

The dynamics of how liability lawsuits are handled emphasize this. The outcome is decided either by a judge (a bench trial) or by a jury. Judges are never engineers, and most lawyers for the injured parties do their best to keep engineers from serving on product liability juries. So, the technical merit of a manufacturing company's position doesn't have the significance that other engineers might be able to see or understand.

All too often, it's really about credibility. If an engineer is called to testify in a product liability trial, the lawyers for the buyer will almost always try and show that the engineer is an idiot. He or she may be brilliant and stand on firm technical ground in testifying, but all that can go out the window if an aggressive lawyer can raise doubt about an engineer's intent or competence.

For anyone who has never been called to testify or give a deposition in a product liability lawsuit, I would like to tell you that an experienced product liability lawyer is very skillful at forcing engineers to become frustrated and angry. Good trial lawyers can often elicit seemingly innocent comments that can then be expanded into a suggestion that you really don't know what you are talking about—even when you do.

---

4. For those who are not familiar with the U.S. system, there's one other aspect of the U.S. system that really changes the landscape in tort law—and is far different than British law. In the United States, both sides pay fees for lawyers and court costs regardless of the outcome of the trial. In most countries, the loser pays both sets of fees. The U.S. approach is based on the idea that a "little guy" can't afford to risk paying for the lawyers of a big company, but this leads to many personal-injury law firms that engage in "recruiting" injured parties for lawsuits.

While there are penalties for frivolous lawsuits, U.S. courts are full of suits with limited merits because attorneys file liability suits on contingency; they tell clients (correctly, as long as this is agreed to between the law firm and the client) that they need not pay anything for legal help unless they win the suit. The calculus is cold but legal and, for the law profession, ethical. File enough suits and win some percentage of them; you'll earn substantial sums on each victorious suit and lose some money on failed suits. The more you file, the greater the difference between what you win and what you lose. So, trial lawyers try to file as many non-frivolous suits as possible and make more money from the way the system works.

This sets the stage for the ethical underpinnings that everyone involved in preparing a DFMEA needs to comprehend. The best policy is to tell the truth. Don't overstate or understate issues; just enter the things that make sense and are, to the greatest extent possible, supported by facts and buttressed by your professional training, experience, and judgment.

The way to think about this is to imagine that a flaw in a product you've worked on has led to a death. You are subpoenaed to testify in the resulting product liability lawsuit. The DFMEA was also subpoenaed,[5] and an expert witness for the victim has studied it in detail.

As a result, the lawyer who will question you is keenly aware of what was and was not on the DFMEA and has a very good idea of how the DFMEA relates to the death.

What do you want to say under oath when you are facing the family of the person who was killed?

A. We didn't say anything about that flaw because we didn't want to end up in court.

B. I don't know why management didn't react to this problem; we all knew it was there, but we changed the wording on the DFMEA to avoid legal problems.

C. We didn't know anything about that flaw. Sorry, but engineering is an uncertain art.

D. We identified that flaw and took the corrective actions shown on the DFMEA worksheet. We made a strong and constructive effort to prevent our product from injuring anyone. Our DFMEA shows that we understood there was risk, and it even shows the extra effort we made to reduce the risks. I am terribly sorry that things turned out the way they did.

If you answered anything but "D" you might be ethically challenged and I can't help you much. "C" might be acceptable if that were the truth, but if you follow the outline in this book, that explanation is unlikely to be true.

The final word in this area is simple. Almost no engineer working on an actual problem is a qualified product liability attorney. Don't try to be

---

5. You can't keep this a secret in a civil trial; the plaintiff's power of discovery—their ability to ask for any and all documents relating to the case—is very broad. A DFMEA is virtually certainly something you would have to hand over, and any attempt to hide this can mean even more trouble.

cunning or believe you can outsmart good lawyers. Do what you do best and use the DFMEA process to prevent flaws in the first place. If things go wrong—and they can, no matter how hard you try—at least you will have the cold comfort of knowing that you have behaved in a moral and ethical fashion.

One final note about product liability law must be addressed. The tables that you use for DFMEA could present a problem, particularly if you use nonstandard tables. This could cause another difficult round of questions for you in testimony.

"What made you so smart that you thought you could use a different set of tables than the rest of your industry?" A lawyer asking this doesn't really care much about the answer you give. Almost any answer will diminish your credibility. Some companies have developed supplemental tables that explain how they will use standard tables; that's not a problem. In fact, it's a good idea. But modifying widely accepted tables or going against industry standards could create difficulties in a product liability lawsuit.

"LIVING DOCUMENT" ISSUES

Often, the term "living document" is used when DFMEA is discussed. What does this mean? To some, it means that a DFMEA must be updated when things change. To others, it means that DFMEA provides a record of engineering and quality efforts in the design process, and such records must be kept up to date. To others, it provides an important history of lessons learned that can be applied to future projects.

All of these ideas are valuable and make it important to use DFMEA results for more than an entry on a checklist.

Updating DFMEA Worksheets

After Step 7 is completed, you have several columns that still require entries before the DFMEA worksheet is complete. We've discussed reporting action plan results previously, and we briefly talked about "final numbers" in Chapter 9.

When the design is ready for release, and all of the recommended actions have been completed, you are ready to enter the final numbers for severity, occurrence, detection, and risk priority on the worksheet. These are the far-right columns on the worksheet, and these numbers should always represent the state of risk you believe exists for a product that has entered commercial production.

These numbers can be different than the numbers that were entered while you were considering the risk inherent in the concept design. In fact, some of the numbers should be different. It's even possible that a classification entry, particularly a "significant" entry, can be eliminated.

If your team didn't make any changes as a result of your work, you really ought to review the worksheet. After all, every new product project ought to yield improvement. And with improvement comes uncertainty, which, in turn, means that some risk is present.

However, any changes in severity, occurrence, or detection from your original estimates must be supported by data, and that data must be readily available as proof of your conclusions about risk.

Beyond completing the worksheet, when you have passed the engineering release milestone, it still remains possible that you will need to make design changes. Every time a change is made, it almost certainly means that the DFMEA, as well as supporting documents such as the block diagram set, the P-diagram, and the list of special characteristics, needs updating.

Different organizations have different policies and procedures for updating the worksheet. On most worksheet forms, the header section acknowledges that changes are possible, with spaces for revision identification and dates.

The "best practice" in this area is to treat changes to the worksheet approximately the same way you might treat a change in an engineering drawing.[6] Record the new information but don't delete the old information, thereby showing the progression of knowledge and understanding that has occurred over time.

Creating a DFMEA Template Library

Another constructive use of DFMEA is to use previous DFMEA studies as templates for next-generation designs. This is a great idea; not every DFMEA has to be done from scratch, and this is particularly true when you are working on an evolutionary project that is only a bit different from a previous tried-and-true project.

Ultimately, a good engineering organization will have a well-structured and frequently accessed library of DFMEA templates that can be used to further reduce the time and effort needed to carefully and fully address a new design. This can be a great time-saver and can also be extremely effective, but you must keep these limitations in mind:

---

6. You don't need all of the details of a typical engineering change order— although updating the DFMEA worksheet really ought to be part of a solid engineering change procedure.

- Using a bad DFMEA study as a template is worse than not doing a DFMEA at all. Using a weak, poorly conceived, or nondeductive DFMEA study as a design baseline gives the impression that a serious study was done. This can lead to overconfidence and rushing of a design to release, an approach that can have devastating consequences.

- What looks the same in generation-to-generation designs is often not really the same. Differences can be significant even if they appear to be superficial. If you're really proficient with the deductive approach, creating a new DFMEA, or at least important sections of a DFMEA, won't be that much work. I've seen many major problems with so-called "carryover" designs—designs that worked fine for years but seemed to develop major problems when used in a slightly different application or changing market conditions.

- DFMEA studies that aren't kept up to date don't form the basis for a sound template. If you leave errors in a previous study and that study is used as a template, you'll simply repeat the old errors. Worse, you'll be completely unaware that you've done this.

If you keep these ideas clearly in mind, a library of templates is a powerful and useful tool for product engineering activities.

Using DFMEA As an Aid in Problem Solving

You may not realize this, but the underlying logic of FMEA and formal problem resolution is virtually identical. To solve a problem in a rigorous manner, most recognized problem-solving methodologies are based on this logical chain:

1. Identify and record the problem. You'll be told about something that was sensed, so the initial problem report, no matter what format it comes in or how it is communicated, will likely correlate with one or more effects on the DFMEA worksheet.

2. Determine the underlying root cause or causal factors that led to the problem. That's something that should be on the DFMEA, too.

3. Determine a specific action to eliminate the cause—or, more realistically, to reduce the rate or frequency at which the cause leads to the problem. If the issue wasn't addressed during the initial development project, you will need to look at the

appropriate function lines on the worksheet and see what you did. Then, you need to formulate a recommended action, just as you would during development.

4. Develop effective controls to ensure that the proposed action is effective. Once again, this is something you should understand very well.

The similarity to the column-by-column approach you've learned in this book should be obvious. In many ways, problem solving is just an effort to repair the errors you made in an earlier FMEA study.

FINAL THOUGHTS

Over many years of effort and learning, I've come to recognize DFMEA as the most important intellectual tool that can be used to assess and understand a new design. But I've also learned that doing this kind of study in a way that is actually helpful, and not just a mind-numbing waste of time, isn't easy.

I've tried to show you many examples of industrial errors that led to terrible outcomes, where people were killed, incredible sums were squandered, and innumerable lives were disrupted. Whether you consider the problems Ford had with the Pinto and Explorer, or NASA with the Apollo I project or the *Columbia* disaster, or the Airbus 380 mess, there was a deficit in understanding critical technical issues that led to very bad or even tragic results.

On the other hand, Orville and Wilbur Wright cheated death. Wilbur died prematurely and tragically from typhoid fever in 1912 at the age of forty-five, while Orville lived a full life, passing away at age seventy-six. But the Wright brothers were the most skillful practitioners of prevention-oriented analytical engineering of their time. Their accomplishments are still remarkable—and they did it all by applying hard and relentless effort to analyze their designs.

For almost two decades I've spent almost all of my professional life teaching people, particularly engineers and technical managers, how to do things in a more structured and productive manner. DFMEA is just one of the techniques I've taught, but it is the single methodology that has fascinated me more than any other. It's rare that I will lead or facilitate a DFMEA session and not learn something new and useful about the technique. And I had ten years of on-again, off-again experience with DFMEA before I started trying to teach people how to use it constructively, so I'm still learning things after more than 30 years of exploration.

I'm sure that you will find the ideas I've presented useful, but I'm just as sure that you'll have additional insights if you practice these ideas. The notion that DFMEA can be more deductive than inductive is powerful, but it does take time and effort to see and harness that power. I hope that you will see that as you use DFMEA in your design efforts.

Bibliography

Chapter 1

Birsch, Douglas, and John Fielder. *The Ford Pinto Case: A Study in Applied Ethics, Business, and Society.* Albany, NY: State University of New York Press, 1994.

Flottau, Jens, and Adrian Schofield. "Airbus A380 Delayed by One More Year; Restructuring Planned." *Aviation Week* (Oct 4, 2006).

National Aeronautics and Space Administration. "Procedure for Failure Mode, Effects and Criticality Analysis." Springfield, VA: National Technical Information Service, 1966.

Peterson, Barbara S. "Airbus A380: Inside Aviation's New Jumbo Trouble." *Popular Mechanics* (October 1, 2009). Available at http://www. popularmechanics.com/science/space/4201627. Accessed 6/7/10.

U.S. Department of Defense. "Procedures for Performing a Failure Mode, Effects and Criticality Analysis." Lakehurst, NJ: Naval Air Engineering Center, Systems Engineering Standardization Department, 1983.

A detailed speech on October 3, 2006 by Christian Streiff about the A380 design and production problem was published in full by a number of internet bloggers even though EADS, Airbus's parent company, did not release the speech (for example, see http://blog.seattlepi.com/aerospace/archives/107302. asp)

Chapter 2

Crouch, Tom. *The Bishop's Boys: A Life of Wilbur and Orville Wright.* New York: W. W. Norton & Company, 1989.

Jakab, Peter L. *Visions of a Flying Machine: The Wright Brothers and the Process of Invention.* Smithsonian History of Aviation and Spaceflight Series. Washington, D.C.: Smithsonian Books, 1997.

Padfield, G. D., and B. Lawrence. "The Birth of Flight Control: An Engineering Analysis of the Wright Brothers' 1902 Glider." *The Aeronautical Journal* (December 2003).

Report of Apollo 204 Review Board, NASA Historical Reference Collection, NASA History Office, 5 April 1967, NASA Headquarters, Washington, DC.

In 2003, the centennial of the first flight, a number of compelling video presentations aired about the Wright Brothers. One episode of the PBS show *Nova* that tracked an effort to replicate the first Wright Flyer is available online at http://www.fancast.com/tv/Nova/9993/770190260/Wright-Brothers%92-Flying-Machine/videos.

Chapter 3

Roger Slater. *Integrated Process Management.* New York: McGraw-Hill, 1991.

Chapter 4

Bartolomei, J. E., and T. Miller. "Functional Analysis Systems Technique (F.A.S.T.) As a Group Knowledge Elicitation Method for Model Building." Report. U.S. Air Force, Wright-Patterson AFB, OH. No date.

Miles, Lawrence D. *Techniques of Value Analysis and Engineering,* 3rd ed. Lawrence D. Miles Value Foundation, 1989.

Chapters 5–10

Automotive Industry Action Group. *Potential Failure Mode and Effects Analysis Reference Manual,* 4th ed. Southfield, MI: AIAG, 2008.

Automotive Industry Action Group. *Production Part Approval Process Reference Manual,* 4th ed. Southfield, MI: AIAG, 2006.

Csere, Csaba. "Why Are Ford Explorers Crashing?" *Car and Driver* (January 2001).

National Aeronautics and Space Administration. "Columbia Accident Investigation Board, Final Report." Washington, D.C.: National Aeronautics and Space Administration and the Government Printing Office, 2003.

National Transportation Safety Board. "In-Flight Breakup over the Atlantic Ocean. Trans World Airlines Flight 800 Boeing 747-131, N93119, Near East Moriches, New York, July 17, 1996." Report.

Sawyer, Kathy. "Witness Chides NASA in Assessing Risk: Ex-Space Official Cites Review of Problem-Plagued 1999 Columbia Mission." *The Washington Post* (March 7, 2003).

There are dozens of authoritative Web sites that describe the McDonald's "hot coffee" lawsuit in detail; lectlaw.com/files/cur78.htm provides a sound summary that has limited bias.

Index